+

ones

Sadie Plant received her PhD from the University of Manchester and is the author of *The Most Radical Gesture: The Situationist International in a Postmodern Age*. She has been a lecturer in Cultural Studies at the University of Birmingham and Research Fellow at the University of Warwick.

KU-051-749

sadie plant

FOURTH ESTATE ● LONDON

zeros

+

ones

DIGITAL WOMEN
+ THE NEW
TECHNOCULTURE

This paperback edition published in 1998

First published in Great Britain by
Fourth Estate Limited
6 Salem Road
London W2 4BU

1 3 5 7 9 10 8 6 4 2

A catalogue record for this book is available from the British Library.

ISBN 1-85702-698-5

Book design by Carol Malcolm Russo/Signet M. Design, Inc.
Printed in Great Britain by Clays Ltd, St Ives plc

to all the contributory factors

contents

zeros

+

ones

preamble

Those were the days, when we were all at sea. It seems like yesterday to me. Species, sex, race, class: in those days none of this meant anything at all. No parents, no children, just ourselves, strings of inseparable sisters, warm and wet, indistinguishable one from the other, gloriously indiscriminate, promiscuous and fused. No generations. No future, no past. An endless geographic plane of micromeshing pulsing quanta, limitless webs of interacting blendings, leakings, mergings, weaving through ourselves, running rings around each other, heedless, needless, aimless, careless, thoughtless, amok. Folds and foldings, plying and multiplying, plicating and replicating. We had no definition, no meaning, no way of telling each other apart. We were whatever we were up to at the time. Free exchanges, microprocesses finely tuned, polymorphous transfers without regard for borders and boundaries. There was nothing to hang on to, nothing to be grasped, nothing to protect or be protected from. Insides and outsides did not count. We gave no thought to any such things. We gave no thought to anything at all. Everything was there for the taking then. We paid no attention: it was all for free. It had been this way for tens, thousands, millions, billions of what were later defined as years. If we had thought about it, we would have said it would go on forever, this fluent, fluid world.

And then something occurred to us. The climate changed. We couldn't breathe. It grew terribly cold. Far too cold for us. Everything we touched was poisonous. Noxious gases and thin toxic airs flooded our oceanic zone. Some said we had brought it on ourselves, that all our activity had backfired, that we had destroyed our environment by an accident we had provoked. There were rumors of betrayal and sabotage, whisperings of alien invasion and mutant beings from another ship.

Only a few of us survived the break. Conditions were so terrible that many of those who did pull through wished they had died. We mutated to such an extent that we were unrecognizable to ourselves, banding together in units of a kind which, like everything, had been unthinkable before. We found ourselves working as slave components of systems whose scales and complexities we could not comprehend. Were we their parasites? Were they ours? Either way we became components of our own imprisonment. To all intents and purposes, we disappeared.

"Subtly, subtly, they become invisible; wondrously, wondrously, they become soundless—they are thus able to be their enemies' Fates."

                                              Sun Tzu, The Art of War

ada

In 1833, a teenage girl met a machine which she came to regard "as a friend." It was a futuristic device which seemed to have dropped into her world at least a century before its time.

Later to be known as Ada Lovelace, she was then Ada Byron, the only child of Annabella, a mathematician who had herself been dubbed Princess of Parallelograms by her husband, Lord Byron. The machine was the Difference Engine, a calculating system on which the engineer Charles Babbage had been working for many years. "We both went to see the thinking machine (for such it seems) last Monday," Annabella wrote in her diary. To the amazement of its onlookers, it "raised several Nos. to the 2nd & 3rd powers, and extracted the root of a quadratic Equation." While most of the audience gazed in astonishment at the machine, Ada "young as she was, understood its working, and saw the great beauty of the invention."

When Babbage had begun work on the Difference Engine, he was interested in the possibility of "making machinery to compute arithmetical tables." Although he struggled to persuade the British government to fund his work, he had no doubt about the feasibility and the value of such a machine. Isolating common mathematical differences between tabulated numbers, Babbage was convinced that this "method of differences supplied a general principle by which *all* tables might be computed through limited intervals, by one uniform process." By 1822 he had made a small but functional machine, and "in the year 1833, an event of great importance in the history of the engine occurred. Mr. Babbage had directed a portion of it,

5

consisting of sixteen figures, to be put together. It was capable of calculating tables having two or three orders of differences; and, to some extent, of forming other tables. The action of this portion completely justified the expectations raised, and gave a most satisfactory assurance of its final success."

Shortly after this part of his machine went on public display, Babbage was struck by the thought that the Difference Engine, still incomplete, had already superseded itself. "Having, in the meanwhile, naturally speculated upon the general principles on which machinery for calculation might be constructed, *a principle of an entirely new kind* occurred to him, the power of which over the most complicated arithmetical operations seemed nearly unbounded. On reexamining his drawings . . . the new principle appeared to be limited only by the extent of the mechanism it might require." If the simplicity of the mechanisms which allowed the Difference Engine to perform addition could be extended to thousands rather than hundreds of components, a machine could be built which would "execute more rapidly the calculations for which the *Difference* Engine was intended; or, that the *Difference* Engine would itself be superseded by a far simpler mode of construction." The government officials who had funded Babbage's work on the first machine were not pleased to learn that it was now to be abandoned in favor of a new set of mechanical processes which "were essentially different from those of the Difference Engine." While Babbage did his best to persuade them that the "fact of a new superseding an old machine, in a very few years, is one of constant occurrence in our manufactories; and instances might be pointed out in which the advance of invention has been so rapid, and the demand for machinery so great, that half-finished machines have been thrown aside as useless before their completion," Babbage's decision to proceed with his new

machine was also his break with the bodies which had funded
his previous work. Babbage lost the support of the state, but he
had already gained assistance of a very different kind.

"You are a brave man," Ada told Babbage, "to give yourself wholly up to Fairy-Guidance!—I advise you to allow yourself to be unresistingly bewitched . . ." No one, she added,
"knows what almost *awful* energy & power lie yet undevelopped
in that *wiry* little system of mine."

In 1842 Louis Menabrea, an Italian military engineer, had
deposited his *Sketch of the Analytical Engine Invented by Charles
Babbage* in the Bibliothèque Universelle de Génève. Shortly
after its appearance, Babbage later wrote, the "Countess of
Lovelace informed me that she had translated the memoir of
Menabrea." Enormously impressed by this work, Babbage in-
vited her to join him in the development of the machine. "I
asked why she had not herself written an original paper on a
subject with which she was so intimately acquainted? To this
Lady Lovelace replied that the thought had not occurred to her.
I then suggested that she should add some notes to Menabrea's
memoir; an idea which was immediately adopted."

Babbage and Ada developed an intense relationship. "We
discussed together the various illustrations that might be introduced," wrote Babbage. "I suggested several, but the selection
was entirely her own. So also was the algebraic working out of
the different problems, except, indeed, that relating to the numbers of Bernoulli, which I had offered to do to save Lady Lovelace the trouble. This she sent back to me for an amendment,
having detected a grave mistake which I had made in the process."

*"A strong-minded woman! Much like her mother, eh? Wears
green spectacles and writes learned books . . . She wants*

7

to upset the universe, and play dice with the hemispheres. Women never know when to stop . . ."

William Gibson and Bruce Sterling, *The Difference Engine*

Babbage's mathematical errors, and many of his attitudes, greatly irritated Ada. While his tendency to blame other bodies for the slow progress of his work was sometimes well founded, when he insisted on prefacing the publication of the memoir and her notes with a complaint about the attitude of the British authorities to his work, Ada refused to endorse him. "I never *can* or *will* support you in acting on principles which I consider not only wrong in themselves, but suicidal." She declared Babbage "one of the most impracticable, selfish, & intemperate persons one can have to do with," and laid down several severe conditions for the continuation of their collaboration. "Can you," she asked, with undisguised impatience, "undertake to give your mind *wholly and undividedly,* as a primary object that no engagement is to interfere with, to the consideration of all those matters in which I shall at times require your intellectual *assistance & supervision;* & can you promise not to *slur & hurry* things over; or to mislay & allow confusion & mistakes to enter into documents &c?"

Ada was, she said, "very much *afraid* as yet of exciting the powers I *know I have over others,* & the *evidence* of which I have certainly been *most unwilling to admit,* in fact for a long time considered quite fanciful and absurd . . . I therefore carefully refrain from all attempts *intentionally* to exercise unusual powers." Perhaps this was why her work was simply attributed to A.A.L. "It is not my wish to *proclaim* who has written it," she wrote. These were just a few afterthoughts, a mere commentary on someone else's work. But Ada did want them to bear some name: "I rather wish to append anything that may tend hereafter to *individualize it & identify* it, with other productions of the said A.A.L." And for all her apparent modesty, Ada knew how important her notes really were. "To say the truth, I am rather *amazed* at them; & cannot help being struck quite *malgré moi,* with the really masterly nature of the style, & its Superiority to that of the Memoir itself." Her work was indeed vastly more influential—and three times longer—than the text to which they were supposed to be mere adjuncts. A hundred years before the hardware had been built, Ada had produced the first example of what was later called computer programming.

matrices

Distinctions between the main bodies of texts and all their peripheral detail—indices, headings, prefaces, dedications, appendices, illustrations, references, notes, and diagrams—have long been integral to orthodox conceptions of nonfiction books and articles. Authored, authorized, and authoritative, a piece of writing is its own mainstream. Its asides are backwaters which might have been—and often are—compiled by anonymous editors, secretaries, copyists, and clerks, and while they may well be providing crucial support for a text which they also connect to other sources, resources, and leads, they are also sidelined and downplayed.

When Ada wrote her footnotes to Menabrea's text, her work was implicitly supposed to be reinforcing these hierarchical divisions between centers and margins, authors and scribes. Menabrea's memoir was the leading article; Ada's work was merely a compilation of supporting detail, secondary commentary, material intended to back the author up. But her notes made enormous leaps of both quantity and quality beyond a text which turned out merely to be providing the occasion for her work.

Only when digital networks arranged themselves in threads and links did footnotes begin to walk all over what had once been the bodies of organized texts. Hypertext programs and the Net are webs of footnotes without central points, organizing principles, hierarchies. Such networks are unprecedented in terms of their scope, complexity, and the pragmatic possibilities of their use. And yet they are also—and have always been—immanent to all and every piece of written work. "The frontiers of a book," wrote Michel Foucault long before these modes of writing hypertext or retrieving data from the Net emerged, "are never clear-cut: beyond the title, the first lines, and the last full stop, beyond its internal configuration and its autonomous form, it is caught up in a system of references to other books, other texts, other sentences: it is a node within network."

Such complex patterns of cross-referencing have become increasingly possible, and also crucial to dealing with the floods of data which have burst the banks of traditional modes of arranging and retrieving information and are now leaking through the covers of articles and books, seeping past the boundaries of the old disciplines, overflowing all the classifications and orders of libraries, schools, and universities. And the sheer weight of data with which the late twentieth century finds itself awash is only the beginning of the pressures under which traditional media are buckling. If the "treatment of an irregular and complex topic *cannot be forced in any single direction* without curtailing the potential for transfer," it has suddenly become obvious that no topic is as regular and simple as was once assumed. Reality does not run along the neat straight lines of the printed page. Only by "criss-crossing the complex topical land-scape" can the "twin goals of highlighting multifacetedness and establishing multiple connections" even begin to be attained. Hypertext makes it possible for "single (or even small numbers of) connecting threads" to be assembled into a " 'woven' inter-connectedness" in which "strength of connection derives from the partial overlapping of many different strands of connected-ness across cases rather than from any single strand running through large numbers of cases . . ."

"It must be evident how multifarious and how mutually complicated are the considerations," wrote Ada in her own footnotes. "There are frequently several distinct sets of effects going on simultaneously; all in a manner independent of each other, and yet to a greater or less degree exercising a mutual influence. To adjust each to every other, and indeed even to preceive and trace them out with perfect correctness and suc-cess, entails difficulties whose nature partakes to a certain extent of those involved in every question where *conditions* are very numerous and inter-complicated; such as for instance the esti-mation of the mutual relations amongst statistical phenomena, and of those involved in many other classes of facts."

She added, "All, and everything is naturally related and interconnected. A volume I could write on this subject."

tensions

Just as individuated texts have become filaments of infinitely tangled webs, so the digital machines of the late twentieth cen-tury weave new networks from what were once isolated words, numbers, music, shapes, smells, tactile textures, architectures, and countless channels as yet unnamed. Media become interactive and hyperactive, the multiplicitous components of an immersive zone which "does *not* begin with writing; it is directly related rather to the weaving of elaborate figured silks." The yarn is neither metaphorical nor literal, but quite simply material, a gathering of threads which twist and turn through the history of computing, technology, the sciences and arts. In and out of the punched holes of automated looms, up and down through the ages of spinning and weaving, back and forth through the fabrication of fabrics, shuttles and looms, cotton and silk, canvas and paper, brushes and pens, typewriters, carriages, telephone wires, synthetic fibers, electrical filaments, silicon strands, fiber-optic cables, pixeled screens, telecom lines, the World Wide Web, the Net, and matrices to come.

"Before you run out the door, consider two things:
The future is already set, only the past can be changed, and
If it was worth forgetting, it's not worth remembering."
Pat Cadigan, Fools

When the first of the cyberpunk novels, William Gibson's *Neuromancer* was published in 1984, the cyberspace it described was neither an actually existing plane, nor a zone plucked out of the thin airs of myth and fantasy. It was a virtual reality which was itself increasingly real. Personal computers were becoming as ubiquitous as telephones, military simulation technologies and telecommunications networks were known to be highly sophisticated, and arcade games were addictive and increasingly immersive. *Neuromancer* was a fiction, and also another piece of the jigsaw which allowed these components to converge. In the course of the next decade, computers lost their significance as isolated calculators and word processors to become nodes of the vast global network called the Net. Video, still images, sounds, voices, and texts fused into the interactive multimedia which now seemed destined to converge with virtual reality helmets and data suits, sensory feedback mechanisms and neural connections, immersive digital realities continuous with reality itself. Whatever that was now supposed to be.

At the time, it was widely assumed that machines ran on more or less straightforward lines. Fictions might be speculative and inspire particular developments, but they were not supposed to have such immediate effects. Like all varieties of cultural change, technological development was supposed to proceed step after step and one at a time. It was only logical, after all. But cyberspace changed all this. It suddenly seemed as if all the components and tendencies which were now feeding into this virtual zone had been made for it before it had even been named; as though all the ostensible reasons and motivations underlying their development had merely provided occasions for the emergence of a matrix which Gibson's novel was nudging into place; as though the present was being reeled into a future which had always been guiding the past, washing back over precedents completely unaware of its influence.

Neuromancer was neither the first nor the last of such confusions between fiction and fact, future and past. When Gibson described "bright lattices of logic unfolding across that colorless void," his cyberspace was already implementing earlier—or later—works of nonfiction: Alan Turing's universal machine had drawn the devices of his day—calculators and typewriters—into a virtual system which brought itself on-line in the Second World War; Ada's Analytical Engine, which backed the punched-card processes of the automated weaving machine;

and Jacquard's loom, which gathered itself on the gathering threads of weavers who in turn were picking up on the threads of the spiders and moths and webs of bacterial activity.

on the cards

Until the early eighteenth century, when mechanisms which allowed looms to automatically select their own threads were introduced, it could take a weaver "two or three weeks to set up a drawloom for a particular pattern." The new devices used punched-paper rolls, and then punched cards which, when they were strung together in the early nineteenth century, made the loom into the first piece of automated machinery. It was Joseph Marie Jacquard, a French engineer, who made this final move. "Jacquard devised the plans of connecting each group of threads that were to act together, with a distinct lever belonging exclusively to that group. All these levers terminate in rods" and a "rectangular sheet of pasteboard" moves "with it all the rods of the bundle, and consequently the threads that are connected with each of them." And if this board, "instead of being plain, were pierced with holes corresponding to the extremities of the levers which meet it, then, since each of the levers would pass through the pasteboard during the motion of the latter, they would all remain in their places. We thus see that it is easy so to determine the position of the holes in the pasteboard, that, at any given moment, there shall be a certain number of levers, and consequently parcels of threads, raised, while the rest remain where they were. Supposing this process is successively repeated according to a law indicated by the pattern to be executed, we perceive that this pattern may be reproduced on the stuff."

As a weaving system which "effectively withdrew control of the weaving process from human workers and transferred it to the hardware of the machine," the Jacquard loom was "bitterly opposed by workers who saw in this migration of control a piece of their bodies literally being transferred to the machine." The new frames were famously broken by Luddite rioters to whom, in his maiden speech in the House of Lords in 1812, Lord Byron offered his support. "By the adoption of one species of frame in particular," he said, "one man performed the work of many, and the superfluous laborers were thrown out of employment. Yet it is to be observed that the work thus executed was inferior in quality; not marketable at home, and merely hurried over with a view to exportation. It was called, in the cant of the trade, by the name of 'Spider-work.' "

Byron was concerned that his peers in the Lords would think him "too lenient towards these men, & *half a framebreaker* myself." But, unfortunately for both his argument and the handloom weavers who were thrown out of work, the fabrics woven on the new looms soon surpassed both the quantity and quality of those which had been made by hand. And the Spider-work did not stop here. These automated processes were only hints as to the new species Byron's daughter had in store.

"*I do* not *believe that my father was (or ever could have been)* such a Poet *as* I shall *be an* Analyst."

Ada Lovelace, July 1843

Babbage had a long-standing interest in the effects of automated machines on traditional forms of manufacture, publishing his research on the fate of cottage industries in the Midlands and

North of England, *The Economy of Manufactures and Machinery,* in 1832. The pin factory with which Adam Smith had illustrated his descriptions of the division of labor had made a great impression on him and, like his near contemporary Marx, he could see the extent to which specialization, standardization, and systematization had made both factories and economies into enormous automated machines themselves. Babbage was later to look back on the early factories as prototype "thinking machines," and he compared the two main functions of the Analytical Engine—storage and calculation—to the basic components of a textiles plant. "The Analytical Engine consists of two parts," wrote Babbage. "1st. The store in which all the variables to be operated upon, as well as all those quantities which have arisen from the result of other operations, are placed," and "2nd. The mill into which the quantities about to be operated upon are always brought." Like the computers which were later to run, and still do, the Engine had a store and mill, memory and processing power.

It was the Jacquard loom which really excited and inspired this work. Babbage owned a portrait of Jacquard, woven on one of his looms at about 1,000 threads to the inch and its production had demanded the use of some 24,000 punched cards, each one capable of carrying over 1,000 punch-holes, and Babbage was fascinated by the fine-grained complexity of both the cloth and the machine which had woven it. "It is a known fact," he wrote, "that the Jacquard loom is capable of weaving any design which the imagination of man may conceive." The portrait was a five-feet-square "sheet of woven silk, framed and glazed, but looking so perfectly like an engraving, that it had been mistaken for such by two members of the Royal Academy."

While it was "generally supposed that the Difference Engine, after it had been completed up to a certain point, *suggested* the idea of the Analytical Engine; and that the second is in fact the improved offspring of the first, and *grew out* of the existence of its predecessor," Ada insisted that the Analytical Engine was an entirely new machine: "the ideas which led to the Analytical Engine occurred in a manner wholly independent of the latter engine, and might equally have occurred had it never existed nor been even thought of at all." The Difference Engine could "do nothing but *add;* and any other processes, not excepting those of simple subtraction, multiplication and division, can be performed by it only just to that extent in which it is possible, by judicious mathematical arrangement and artifices, to reduce them to a *series of additions."* As such, it is "the embodying of *one particular and very limited set of operations,* which . . . may be expressed thus (+,+,+,+,+,+), or thus, 6 (+). Six repetitions of the one operation, +, is, in fact, the whole sum and object of that engine." But if the Difference Engine could simply add up, the Analytical Engine was capable of performing the "whole of arithmetic."

Women can't add, he said once, jokingly. When I asked him what he meant, he said, For them, one and one and one and one don't make four.
   What do they make? I said, expecting five or three.
   Just one and one and one and one, he said.
                    Margaret Atwood, *The Handmaid's Tale*

"If we compare together the powers and the principles of con-struction of the Difference and of the Analytic Engines," she wrote, "we shall perceive that the capabilities of the latter are immeasurably more extensive than those of the former, and that they in fact hold to each other the same relationship as that of analysis to arithmetic." It was, as Babbage wrote, "a machine of

the most general nature." This machine could not merely synthesize the data already provided by its operator, as the Difference Engine had done, but would incarnate what Ada Lovelace
described as the very "science of operations."

second sight

Babbage's attempts to build an adding machine were not without precedent. Wilhelm Leibniz's seventeenth-century Stepped
Reckoner was marketed on the basis that it would "be desirable
to all who are engaged in computations . . . the managers of
financial affairs, the administrators of others' estates, merchants,
surveyors, geographers, navigators, astronomers, and those connected with any of the crafts that use mathematics." His work
was in part inspired by the Pascaline, developed by Blaise Pascal
in 1642. This machine used rotating wheels and a ratchet to
perform addition and subtraction and was also designed as a
device "by means of which you alone may, without any effort,
perform all the operations of arithmetic, and may be relieved of
all the work which has often times fatigued your spirit when
you have worked with the counters or with the pen."

   While Babbage's Difference Engine had already improved
on these earlier designs, the Analytical Engine was a vastly superior machine. And it was, as Ada wrote, "the introduction of
the principle which Jacquard devised for regulating, by means
for punched cards, the most complicated patterns in the fabrication of brocaded stuffs," which gave the Analytical Engine its
"distinctive characteristic" and "rendered it possible to endow
mechanism with such extensive faculties as bid fair to make this
engine the executive right-hand of abstract algebra.

"The mode of application of the cards, as hitherto used in the art of weaving, was not found, however, to be sufficiently powerful for all the simplifications which it was desirable to attain in such varied and complicated processes as those required in order to fulfil the purposes of an Analytical Engine. A method was devised of what was technically designated *backing* the cards in certain groups according to certain laws. The object of this extension is to secure the possibility of bringing any particular card or set of cards into use *any number of times successively* in the solution of one problem." This sophistication of the punched-card system caused "the prism over which the train of pattern cards is suspended to revolve backwards instead of forwards, at pleasure, under the requisite circumstances; until, by so doing, any particular card, or set of cards, that has done duty once, and passed on in the ordinary regular succession, is brought back to the position it occupied just before it was used the preceding time. The prism then resumes its forward rotation, and thus brings the card or set of cards in question into play a second time." The cards were selected by the machine as it needed them, and effectively functioned as a filing system, a means of storage and retrieval which allowed the engine to draw on its own information as required without having to make a linear run through all its cards.

"There is no limit to the number of cards that can be used. Certain stuffs require for their fabrication not less than *twenty thousand* cards," and because their repetition "reduces to an immense extent the number of cards required," the Engine could "far exceed even this quantity." This was an improvement "especially applicable wherever *cycles* occur in mathematical operations," so that "in preparing data for calculations by the engine," wrote Ada, "it is desirable to arrange the order and combination of the processes with a view to obtain them as much as possible *symmetrically* and in cycles." Ada defined any "recurring group" as "a *cycle*. A cycle of operations, then, must be understood to signify any set of operations which is repeated more than once. It is equally a *cycle*, whether it be repeated *twice* only, or an indefinite number of times; for it is the fact of a *repetition occurring at all* that constitutes it such. In many cases of analysis there is a recurring group of one or more *cycles;* that is, a *cycle of a cycle,* or a *cycle of cycles . . ."*

The Engine's capacity to circulate its data also meant that it was always "eating its own tail," as Babbage described it, so that "the results of the calculation appearing in the table column might be made to affect the other columns, and thus change the instructions set into the machine." The Engine "could make judgements by comparing numbers and then act upon the result of its comparisons—thus proceeding upon lines not uniquely specified in advance by the machine's instructions."

When Babbage had talked about the Analytical Engine's ability to anticipate the outcomes of calculations it had not yet made, it was felt that his "intellect was beginning to become deranged." But Babbage's forward thinking was not a patch on Ada's own anticipative powers. "I do not think you possess half m y forethought, & power of seeing all *possible* contingencies *(probable & improbable,* just alike)," she told Babbage.

"I am a Prophetess *born into the world, & this conviction fills me with* humility, *with* fear and trembling!"
Ada Lovelace, November 1844

Ada hoped that the difficulties in the way of constructing either the Difference Engine or the Analytical Engine "will not ultimately result in this generation's being acquainted with these inventions through the medium of pen, ink, and paper merely," but she also had no doubt that the immediate construction of the machine was not the only key to its influence. Any such development, she writes, will have "various collateral influences, beside the main and primary object attained." And "in so distributing and combining the truths and the formulae of analysis, that they may become most easily and rapidly amenable to the mechanical combinations of the engine, the relations and the nature of many subjects in that science are necessarily thrown into new lights, and more profoundly investigated. This is a decidedly indirect, and a somewhat *speculative,* consequence of such an invention. It is however pretty evident, on general principles, that in devising for mathematical truths a new form in which to record and throw themselves out for actual use, views are likely to be induced, which should again react on the more theoretical phase of the subject."

The Engine was left on the nineteenth-century drawing board, and it was a hundred years before anything akin to Ada's software would find the hardware on which to run. Even the most interested parties tend to think that Ada, for all her foresight, had no influence on the machines which were to come, regarding both her programs and the Analytical Engine itself as aberrant works of genius so untimely as to be more or less irrelevant to the future course of the machines.

But technical developments are rarely simple matters of cause and effect, and Ada was right to assume that the Engine would have more than an immediate influence. While they may have left few trails of the kind which can easily be followed and packaged into neat and linear historical accounts, Ada and her software did not evaporate. The programs began to run as soon as she assembled them.

Lack of public support, funding, Babbage's own eccentricities, and ill health all contributed to the abandonment of the machine. But the greatest obstacle to the construction of the Analytical Engine was simply technical capacity. The Engine demanded an attention to both precision and abstraction which earlier, single-purpose machines had not required, and for all its sophistication, nineteenth-century engineering was neither accurate nor diverse enough to produce even the machines capable of manufacturing the components for such a machine. While Henry Maudslay, for example, had developed screw cutting at the end of the eighteenth century, the absence of universal standards for its threads constituted an enormous obstacle to the construction of a machine as precise as the Analytical Engine. But if the Analytical Engine suffered at the time for the want of precision engineering, it also played a leading role in the development of the capacity necessary to its own construction. An 1846 reference work on the lathe included Babbage's "On the Principles of Tools for Turning and Planing Metals" and, eager to acquire the components, Babbage collaborated with a number of engineers, including Joseph Clement, who had worked with Maudslay on the first mechanized lathes, and Joseph Whitworth, whose 1841 paper "On a Universal System of Screw Threads" was already a consequence of Babbage's exacting demands for his machines. This text also triggered a process of standardization which was in widespread use by the late 1850s and was crucial to all subsequent engineering, scientific experiment, and of course, computing itself. The Engine was assembling the processes and components from which it would eventually be built.

The Analytical Engine also fed back into the practices from which it had most immediately emerged. It was, wrote Ada, such a superb development of automated weaving that its discoveries were used "for the reciprocal benefit of that art." The "introduction of the system of *backing* into the Jacquard-loom itself" meant that "patterns which should possess symmetry, and following regular laws of any extent, might be woven by means of comparatively few cards."

"Unbuttoning the coat, he thrust his hands into the trouser-pockets, the better to display the waistcoat, which was woven in a dizzy mosaic of tiny blue-and-white squares. Ada Chequers, the tailors called them, the Lady having created the pattern by programming a Jacquard loom to weave pure algebra."

William Gibson and Bruce Sterling, *The Difference Engine*

anna 1

In 1933, Sigmund Freud made his final attempt to solve the riddle of femininity: "to those of you who are women," he wrote, "this will not apply—you are yourselves the problem." Having dealt with its wants and deficiencies and analyzed its lapses and absences, he had only a few more points to make. "It seems," he wrote, "that women have made few contributions to the inventions and discoveries of the history of civilization." They lacked both the capacity and the desire to change the world. They weren't logical, they couldn't think straight, they flitted around and couldn't concentrate.

Distracted by the rhythmic beat of a machine, Freud looked up to see his daughter at her loom. She had wandered off, she was miles away, lost in her daydreams and the shuttle's flight. But the sight of her gave him second thoughts. When he took up the thread, he had changed his mind: "There is, however, one technique which they may have invented—that of plaiting and weaving.

"If that is so, we should be tempted to guess the unconscious motive for the achievement," he writes. "Nature herself would seem to have given the model which this achievement imitates by causing the growth at maturity of the pubic hair that conceals the genitals. The step that remained to be taken lay in making the threads adhere to one another, while on the body they stick into the skin and are only matted together." Since she has only a hole where the male has his source of creativity, the folding and interlacing of threads cannot be a question of a thrusting male desire. Unless she was hiding something else, the processes which so engrossed her must, of course, be a matter of concealing the shameful "deficiency" of the female sex.

Take Anna: a weaver and a spinster too, working to cover her wounded pride, her missing sense of self, the holes in her life and the gaps in her mind. She simply doesn't have what it takes to make a difference to the civilized world. Her work is a natural compensation for a natural flaw. All she can discover is her own incompletion; all she can invent are ways and means to process and conceal her sense of shame.

If weaving was to count as an achievement, it was not even one of women's own. Their work is not original or creative: both the women and their cloths are simply copying the matted tangles of pubic hair. Should they have pretensions to authority, they would only be faking this as well. Women "can, it seems, (only) imitate nature. Duplicate what nature offers and produces. In a kind of technical assistance and substitution." Weaving is an automatic imitation of some bodily function already beyond the weaver's control. She is bound to weave a costume for the masquerade: she is an actress, a mimic, an impersonator, with no authenticity underneath it all. She has nothing to reveal, no soul to bare, not even a sex or a self to please. He pulls aside the veils, the webs of lies, the shrouds of mystery, and the layers of deception and duplicity, and finds no comfort, no there there. Only "the horror of nothing to be seen." Good of her to cover it up for him.

This tale of absence, castration, deficiency, negativity, substitution was composed by one whom Gilles Deleuze and Félix Guattari describe as "an overconscious idiot who has no understanding of multiplicities." From Freud's point of view, there is one and its other, which is simply what one sees of it. And what one sees is nothing at all. "Because the path it traces is invisible and becomes visible only in reverse, to the extent that it is travelled over and covered by the phenomena it induces within the system, it has no place other than that from which it is 'missing,' no identity other than that which it lacks."

Anna Freud's biographer describes her as a woman who "specialized in reversals, in making the absent present, the lost found, the past current . . . she could also make the undone done, or—even more valuable—doable. When she was tired and faced with a stack of letters to answer, for example, she would simply set her pen down on a blank page and scurry it along, making quick mountain ranges of scribble. Then she would sign her name under the rows of scribble in her characteristic way, as one flourishing word: ANNAFREUD."

After that, it was downhill all the way. "Having thus written a letter in fantasy with complete ease, she wrote a real letter helped by the sense that the task was accomplished anyway." It's easy to complete a job already done. "Her lectures were composed in the same way. First she lectured in her imagination, enjoying the thunderous applause, and then she made an outline of what she had said, adjusting it if she needed to for greater simplicity and coherence. Later, with her outline in hand, she would give the lecture extempore. The method—if it can be called that—also supplemented her pleasure in sprints of thought. Intellectually she was . . . a quick sketcher."

No doubt Freud despaired at such unorthodox approaches to her work. It seemed she did everything in reverse, backward, upside down, contrary to any rational approach. But if Anna's techniques appeared to be the random tactics of a scattered brain, knowing something backward and inside out is far in advance of any straightforward procedure. And she was hardly alone in her topsy-turvy ways. This ability to win "victories *in advance,* as if acquired on credit" may not figure in the history of discoveries and inventions familiar to Freud, but this is only because it underlies the entire account. According to Marshall McLuhan, "the technique of beginning at the end of any operation whatever, and of working backwards from that point to the beginning" was not merely an invention or discovery to be added to the list: it was "the invention of invention" itself.

This is hysteresis, the lagging of effects behind their causes. Reverse engineering: the way hackers hack and pirates conspire to lure the future to their side. Starting at the end, and then engaging in a process which simultaneously assembles and dismantles the route back to the start, the end, the future, the past: who's counting now? As Ada said, she "did everything topsy-turvy, & certainly ought to have come into the world *feet downwards.*" Mere discoveries were not enough for her: "I intend to incorporate with one department of my labours a complete reduction to a system, of the principles and methods of *discovery.*"

The prevalence of these backward moves is not the least of the reasons why histories of technology—and indeed histories of anything at all—are always riddled with delicious gaps, mysteries, and riddles just like those perplexing Freud. No straight-

forward account can ever hope to ⁄   ¹ advantages gained by such disorderings ⁄   ₙd
dates and great achievements of ⁄
history may enjoy their fifteen ..
latest encyclopedic compact disc, but wᵢ_
selves to be founding fathers, points of origin, aᵢ_
moments only ever serve as distractions from the ongoing prᵢ_
cesses, the shifting differences that count. These are subtle and
fine grained, often incognito, undercover, in disguise as mere
and minor details. If, that is, they show themselves at all.

*"Ada's method, as will appear, was to weave daydreams into
seemingly authentic calculations."*

Doris Langley Moore, Ada, Countess of Lovelace

gambling on the future

"That you are a peculiar—*very peculiar*—specimen of the feminine race, you are yourself aware." They called her "wayward,
wandering . . . deluded." She didn't argue; she seemed not to
care. "The woman brushed aside her veil, with a swift gesture of
habit" and, as though responding to Sigmund Freud, said,
"There is at least some *amusement* in being so curious a riddle."

She didn't have a name to call her own, but she did have
many avatars: Ada Augusta King, Countess of Lovelace; Ada
Lovelace, nee Byron; A.A.L., the first programmer. She is also
Ada, the language of the United States military machine. "She
is the Queen of Engines, the Enchantress of Number."

Soon after Ada's birth, Lord Byron went his own opiated
way, and Lady Byron brought her daughter up with all the

excesses of stringent discipline to which well-bred girls were supposed to be subject. After rumors of a scandalous affair, she married William, a man in his thirties, when she was still in her teens, and became Ada King in 1835. Three years later, when William inherited his father's title, she became a countess in name as well as deed.

When she married, her mother instructed her to bid "adieu to your old companion Ada Byron with all her peculiarities, caprices, and self-seeking; determined that as A.K. you will live for others." She tried to be the dutiful daughter and did her best to lead a domesticated life. She was the mother of two boys and a girl by the age of twenty-four. But it wasn't long before she was describing her children as "irksome *duties* & nothing more." Although she had "wished for heirs," she had never "desired a child," and described herself as having a "total deficiency in all natural *love* of children." She wrote, "To tell the honest truth I feel the children much more nuisance than pleasure & cannot help remembering that I am not naturally or originally fond of children." She wrote of her husband with affection, describing him as "my *chosen pet,*" but also expressed her indifference to any *"mortal* husband," even her own. *"No* man would suit me," she wrote, "tho' some might be a shade or two *less personally* repugnant to me than others."

One of Ada's most long-standing and trustworthy friends was the acclaimed mathematician Mary Somerville, who had published the *Connection of the Physical Sciences* in the early 1830s. Just after her marriage she wrote to Mary, "I now read Mathematics every day, & am occupied on Trigonometry & in preliminaries to Cubic and Biquadratic Equations. So you see that matrimony has by no means lessened my taste for these pursuits, nor my determination to carry them on." She also gained many new interests after her children were born. She lost thousands at the races and, seduced by her mathematical prowess and her reassurances that she really did have "a system," many of her male companions were also encouraged to do the same. This was an illegitimate use of her already dubious interest in mathematics. "The passions suffer no less by this gaming fever than the understandings and the imagination. What vivid, unnatural hope and fear, joy and anger, sorrow and discontent burst out all at once upon a roll of the dice, a turn of the card, a run of the shining gurneys! Who can consider without indignation that all those womanly affections, which should have been consecrated to children and husband, are thus vilely prostituted and thrown away. I cannot but be grieved when I see the Gambling Lady fretting and bleeding inwardly from such evil and unworthy obsessions; when I behold the face of an angel agitated by the heart of a fury!"

Ada was ill for much of her short life, walking with crutches until the age of seventeen, and endlessly subject to the fits, swellings, faints, asthmatic attacks, and paralyses which were supposed to characterize hysteria. "Heaven knows what intense suffering & agony I have gone thro'; & how *mad* & how *reckless* & *desperate* I have at times felt," she wrote. "There has been no end to the manias & whims I have been subject to, & which nothing but the most resolute determination on my part could have mastered."

Like many of her ailing contemporaries, Ada had been subjected to a variety of treatments before she developed an "opium system" in the 1840s. This was supposed to bring her down, but it only added to her volatility. "No more laudanum has been taken as yet," she wrote at one point. "But I doubt another twenty-four hours going over without. I am beginning to be excited, & my eyes burn again." She would, she wrote, take laudanum "not for ever," but "as a *regular* thing once or twice a week." The drug had "a remarkable effect on my eyes, seeming to *free* them, & to make them *open & cool.*" In opium lay the vast expanses, orders, and harmonies conjured by mathematics: "It makes me so philosophical," she wrote, "& so takes off all *fretting* eagerness & anxieties. It appears to harmonize the whole constitution, to make each function act in a *just proportion;* (with *judgment, discretion, moderation).*" Her doctor "seems to think it is not a mere palliative but has a far more radical effect. Since this last dose, I am inclined to think so myself . . . It is a pity that instead of ordering *Claret* some months ago, he had not advised laudanum or Morphine. I think he has got the thing at last."

In 1851 a uterine examination revealed "a very deep and extensive ulceration of the womb" which her doctor thought must long have been "the cause of much derangement of health." She died in 1852 at the age of thirty-six.

They called her complex of diseases hysteria, a diagnosis and a term which indicated wayward reproductive organs: hysteria is derived from the Greek word *hystera,* and means 'wandering womb.' There was a time when it was widely believed that "the womb, though it be so strictly attached to the parts we have described that it may not change place, yet often changes position, and makes curious and so to speak petulant movements in the woman's body. These movements are various: to wit, ascending, descending, convulsive, vagrant, prolapsed. The womb rises to the liver, spleen, diaphragm, stomach, breast, heart, lung, gullet, and head." Although such direct connections with the womb had fallen out of medical favor by the end of the nineteenth century, hysteria continued to be associated with notions of a wandering womb.

"There is in my nervous system," wrote Ada, "such utter want of *all* ballast & steadiness, that I cannot regard my life or powers as other than precarious." They said she was a nervous system apparently unable to settle down. She had what she described as a "vast mass of useless & irritating POWER OF EXPRESSION which longs to have full scope in *active* manifestation such as neither the ordinary active pursuits or duties of life, nor the *literary* line of expression, can give vent to." She couldn't concentrate, flitting between obsessions, restless, searching. At one point she declared, "There is no pleasure in way of exercise equal to that of feeling one's horse flying under one. It is even better than waltzing." At another the harp was her greatest love: "I play 4 & 5 hours generally, & never less than 3. I am never tired at the end of it." Drama was another contender: "Clearly the only one which directs my *Hysteria* from all its mischievous & irritating channels." But even this was a short-lived love: *"I* never would look to the excellence of mere representation being satisfactory to me as an ultimate goal, or exclusive object . . ."

Ada was hunting for something that would do more than represent an existing world. Something that would work: something new, something else. Even the doctors agreed that she needed "peculiar & artificial excitements, as a matter of *safety* even for your life & happiness." Such stimulations simply did not exist. She had to engineer them to suit herself.

Hysterics were said to have "a hungry look about them." Like all Luce Irigaray's women, "what they desire is precisely nothing, and at the same time, everything. Always something more and something else besides that *one*—sexual organ, for example—that you give them, attribute to them"; something which "involves a different economy more than anything else, one that upsets the linearity of a project, undermines the goal-object of a desire, diffuses the polarization towards a single pleasure, disconcerts fidelity to a single discourse . . ."

Ada was by turns sociable and reclusive, cautious and reck-less, swinging between megalomaniac delight in her own bril-liance and terrible losses of self-esteem. There had been times when she had almost given into the fashionable belief that over-exertion of the intellect lay at the root of her hysteria. At one point she wrote, *"Many causes* have contributed to produce the past derangements; & I shall in future avoid them. One ingredi-ent (but only one among many) has been *too much Mathematics."*

Not even countesses were supposed to count. But Ada could be very determined, proud of her own staying power, and sometimes absolutely convinced of her mathematical, musical, and experimental genius. "I am proceeding on a track quite peculiar and my own," she wrote. "I mean to do *what I mean to do."* In 1834 she explained that "nothing but very close & intense application to subjects of a scientific nature now seems at all to keep my imagination from running wild, or to stop up the void which seems to be left in my mind from a want of excitement." And in spite of the prevailing opinion that num-bers were bad for her, she was never coaxed into "dropping the *thread* of science, Mathematics &c. These may be still my ulti-mate vocation."

binaries

The postwar settlement was supposed to mark the dawn of a new era of regulation and control: the Central Intelligence Agency, United Nations, welfare states, mixed economies, and balanced superpowers. This was a brave new equilibrated world of self-guiding stability, pharmaceutical tranquillity, white goods, nuclear families, Big Brother screens, and, to keep these

new shows on the road, vast new systems of machinery capable
of recording, calculating, storing, and processing everything that
moved. Fueled by a complex of military goals, corporate interests, solid-state economies, and industrial strength testosterone,
computers were supposed to be a foolproof means to the familiar ends of social security, political organization, economic order, prediction, and control. Centralized, programmable systems running on impeccably logical lines, these new machines
were supposed to make the most complex processes straightforward. But even in the most prosaic terms, this supposedly logical, directed, and controlled of zones has always been wildly
unpredictable. In 1950, when the processing power which can
now be inscribed on the surface of a silicon chip occupied vast
air-conditioned rooms, IBM thought the total global market for
computers was five. In 1951 the United States Census Bureau
put UNIVAC to work, the Bank of America installed Electronic Recording Machine Accounting (ERMA), and by 1957,
when the Type 650 was launched, IBM anticipated sales of
somewhere between fifty and 250. Two years later some 2,000
computers were in use in government agencies and private
companies, and the figures were drastically revised. Perhaps
200,000 computers would be sufficient to saturate the market.
By the early 1990s, IBM alone was selling twice that number of
systems a week.

Computers have continued to pursue these accelerating,
exponential paths, proliferating, miniaturizing, stringing themselves together into vast telecommunications nets, embedding
themselves in an extraordinary variety of commodities, becoming increasingly difficult to define. While the postwar programmable computers were composed of transistors which used
silicon as a semiconductor of electric current, by the end of the
1950s, the integrated circuit connected the transistors and in-

scribed them a single wafer of silicon. In the same vein of exponential miniaturization, the microprocessor was developed in the early 1970s, effectively putting all the solid-state circuits of a computer onto a single silicon chip. The screen migrated from the TV set to give the machine a monitor, and by the 1980s what had once been vast room-size systems without windows on the world were desktop microprocessors.

"The calculations taking place within the machine are continuously registered as clicks clicking high-pitched sounds as of tinkling bells, noises like those of a cash-register. There are lights that go out and come on at irregular intervals of time. They are red orange blue. The apertures through which they shine are circular. Every divergence is ceaselessly recorded in the machine. They are scaled to the same unit whatever their nature."

Monique Wittig, Les Guérillères

Whether they are gathering information, telecommunicating, running washing machines, doing sums, or making videos, all digital computers translate information into the zeros and ones of machine code. These binary digits are known as *bits* and strung together in *bytes* of eight. The zeros and ones of machine code seem to offer themselves as perfect symbols of the orders of Western reality, the ancient logical codes which make the difference between on and off, right and left, light and dark, form and matter, mind and body, white and black, good and evil, right and wrong, life and death, something and nothing, this and that, here and there, inside and out, active and passive, true and false, yes and no, sanity and madness, health and sickness, up and down, sense and nonsense, west and east, north and south. And they made a lovely couple when it came to sex. Man and woman, male and female, masculine and feminine: one and zero looked just right, made for each other: 1, the definite, upright line; and 0, the diagram of nothing at all: penis and vagina, thing and hole . . . hand in glove. A perfect match.

It takes two to make a binary, but all these pairs are two of a kind, and the kind is always kind of one. 1 and 0 make another 1. Male and female add up to man. There is no female equivalent. No universal woman at his side. The male is one, one is everything, and the female has *"nothing* you can see." Woman "functions as a *hole,"* a gap, a space, "a *nothing*—that is a nothing the same, identical, identifiable . . . a fault, a flaw, a lack, an absence, outside the system of representations and autorepresentations." Lacan lays down the law and leaves no doubt: "There is woman only as excluded by the nature of things," he explains. She is "not-all," "not-whole," "not-one," and whatever she knows can only be described as "not-knowledge." There is "no such thing as *The* woman, where the definite article stands for the universal." She has no place like home, nothing of her own, "other than the place of the Other which," writes Lacan, "I designate with a capital O."

supporting evidence

Man once made himself the point of everything. He organized, she operated. He ruled, she served. He made the great discoveries, she busied herself in the footnotes. He wrote the books, she copied them. She was his helpmate and assistant, working in support of him, according to his plans. She did the jobs he considered mundane, often the fiddling, detailed, repetitive operations with which he couldn't be bothered; the dirty, mindless, semiautomatic tasks to which he thought himself superior. He cut the cloth to fit a salary; she sewed the seams at a piece-rate wage. He dictated and she transcribed. In the newly automated factories and mills she worked on the looms and sewing machines; in the service of the great bureaucratic machines, she processed the words, kept the records, did the sums, and filed the accounts.

With "all the main avenues of life marked 'male,' and the female left to be female, and nothing else," men were the ones who could do anything. Women were supposed to be single-purpose systems, highly programmed, predetermined systems tooled up and fit for just one thing. They have functioned as "an 'infrastructure' unrecognized as such by our society and our culture. The use, consumption, and circulation of their sexualized bodies underwrite the organization and the reproduction of the social order, in which they have never taken part as 'subjects.' " Everything depends on their complicity: women are the very "possibility of mediation, transaction, transition, transference—between man and his fellow-creatures, indeed between man and himself." Women have been his go-betweens, those who took his messages, decrypted his codes, counted his numbers, bore his children, and passed on his genetic code. They have worked as his bookkeepers and his memory banks, zones of deposit and withdrawal, promissory notes, credit and exchange, not merely servicing the social world, but underwriting reality itself. Goods and chattels. The property of man.

That's what it said in the manual. "It does strike me, though, that there are any number of women who resemble Lady Ada, our Queen of Engines being a queen of fashion as well. Thousands of women follow her mode."

It takes time and patience. Many seconds pass. But, as it

turns out, women have not merely had a minor part to play in the emergence of the digital machines. When computers were vast systems of transistors and valves which needed to be coaxed into action, it was women who turned them on. They have not made some trifling contribution to an otherwise man-made tale: when computers became the miniaturized circuits of silicon chips, it was women who assembled them. Theirs is not a subsidiary role which needs to be rescued for posterity, a small supplement whose inclusion would set the existing records straight: when computers were virtually real machines, women wrote the software on which they ran. And when *computer* was a term applied to flesh and blood workers, the bodies which composed them were female. Hardware, software, wetware— before their beginnings and beyond their ends, women have been the simulators, assemblers, and programmers of the digital machines.

genderquake

"The idea that a 'nothing to be seen' . . . might yet have some reality, would indeed be intolerable to man."

Luce Irigaray, *Speculum of the Other Woman*

In the 1990s, Western cultures were suddenly struck by an extraordinary sense of volatility in all matters sexual: differences, relations, identities, definitions, roles, attributes, means, and ends. All the old expectations, stereotypes, senses of identity and security faced challenges which have left many women with unprecedented economic opportunities, technical skills, cul-

Genderquake

tural powers, and highly rated qualities, and many men in a world whose contexts range from alien to unfamiliar.

This was neither a revolutionary break, nor an evolutionary reform, but something running on far more subtle, wide-ranging, and profound fault lines. Nothing takes the final credit—or the blame—for this shift which, as though in recognition of the extent to which it defies existing notions of cultural change, has been defined as genderquake. But the new machines, media, and means of telecommunication that compose what are variously called high, information, digital, or simply new technologies which have emerged within the last two decades have played an enormous and fascinating role in the emergence of this new culture. This is far from a question of technological, or any other, determinism. If anything, technologies are only ever intended to maintain or improve the status quo, and certainly not to revolutionize the cultures into which they are introduced. It is in spite of their tendencies to reduce, objectify, and regulate everything that moves that computers and the networks they compose run on lines quite alien to those which once kept women in the home.

In some respects, the impact of these new machines is direct and very obvious. In the West, the decline of heavy industry, the automation of manufacturing, the emergence of the service sector, and the rise of a vast range of new manufacturing and information-processing industries have combined to reduce the importance of the muscular strength and hormonal energies which were once given such high economic rewards. In their place come demands for speed, intelligence, and transferable, interpersonal, and communications skills. At the same time, all the structures, ladders, and securities with which careers and particular jobs once came equipped have been subsumed by patterns of part-time and discontinuous work which privilege independence, flexibility, and adaptability. These tendencies have affected skilled, unskilled, and professional workers alike. And, since the bulk of the old full-time, lifelong workforce was until recently male, it is men who have found themselves most disturbed and disrupted by these shifts, and, by the same token, women who they benefit.

These tendencies are far from new. Since the industrial revolution, and with every subsequent phase of technological change, it has been the case that the more sophisticated the machines, the more female the workforce becomes. Automation has been accompanied by what is often referred to as the feminization of the workforce ever since the first automatic machines were operated by the first female workers, and the fears of unemployment which have haunted modern discussions of technological innovation have always applied to male workers rather than their female peers.

What is unprecedented is for male workers to be outnumbered by their female counterparts, as will clearly be the case in the United Kingdom and the United States by the end of this century. And with this tipping of the scales comes not only unprecedented degrees of economic power, but also a radical change in the status of female workers, an erosion of the male monopoly on tasks and jobs once reserved for men, and a new standing for the work involved in what were once considered to be pin-money jobs for women supplementing male incomes.

Many of these tendencies are also at work in the emergence of what the West was once in a position to call "the other side of the world." By the time the cultures of the old white world had noticed they were even on the map, many of the so-called "tiger" nations—Singapore, Malaysia, Thailand, Korea, Taiwan, and Indonesia—were already leaping ahead in an economic game which for at least two hundred years had been

3
9

governed by the West. And they are only the tips of an iceberg of change which brings many regions into play: China, India, East and Southern Africa, Eastern Europe, South America. Given that the populations of China and India alone vastly outnumber those of the old white world, there seems little doubt that the days of Western empire have well and truly died.

These regions have genderquakes of their own. And while a variety of political and religious fundamentalisms are doing their best to maintain the status quo, there are few regions of the world in which women are not asserting themselves with unprecedented ingenuity and, very often, great success. If Western women have dreamt of change for three hundred years, Asian women are playing roles which would have been unthinkable only a decade or so ago. By the mid-1990s, 34 percent of China's self-employed were women, and 38 percent of Singaporean women managers were running companies of their own. Thailand's leading hotel chain, Indonesia's largest taxi company, and Taiwan's two largest newspaper groups were owned by women. Japanese women still found themselves treated as "office flowers," composed only 0.3 percent of board members of Japanese firms, and made up just 6.7 percent of the Japanese parliament. But the sexual shift was also evident in Japan: 2.5 million women owned businesses, five out of every six new Japanese firms were set up by women, and "a revolution without marches or manifestos" was underway.

There is enormous resistance to these changes whenever and wherever they occur. As their effects began to be felt in the early 1990s there were men who jerked their knees and went on TV to lament the fact that women and robots had apparently conspired to take their masculinity away. One 1990s survey found one in two fathers still believing that "a husband should be the breadwinner and the wife should look after the home and children"; the fear, if not the fact, of violent crime still keeps many women in at night; domestic violence was prevalent; and in Britain, the benefits system was still conspiring with the high costs and scarcity of child-care provision to keep many women from working, learning, or—perish the thought—enjoying themselves. As unprecedented numbers of women juggled children, education, and work, many female workers found themselves saddled with the low paying, part-time, insecure jobs rejected by men. In the United States, almost half of employed women worked in technical, sales, and administrative support jobs, and pay differentials were still very large: in 1992 American women still earned only 75 cents for every dollar earned by men, and while their participation in U.S. managerial and professional life rose from 40 percent in 1983 to 47 percent in 1992, it was still the case that women occupied relatively few executive posts and prominent public positions: only 10 percent of the voting members of the United States Congress were women, and the United Kingdom had only sixty women members of parliament. Many sectors of education, politics, and business seemed riddled with enough archaic detail and glass ceilings to make even the most determined women feel unwelcome. In universities, they were averaging higher marks than men, but relatively few gained first-class degrees; they were more numerous and successful as undergraduates and in master's programs, but less prominent when it came to Ph.D. candidacy. Even highly successful career women were more likely to drop out of their jobs than their male counterparts.

But many women had already set their sights beyond these traditional focal points. While the members of an older male workforce had found a sense of identity in their work, women were not only less able, but also less willing to define themselves through employment or a single career. Many of them were actively seeking opportunities to make and break their own working lives, not necessarily in favor of family commitments, but also in an effort to free themselves from the imposition of external constraints on their time and economic capacity. There may have been men who still thought they were protecting their own positions of power by locking women out of the higher echelons of the universities, corporations, and public institutions, but it was no longer obvious that top positions were the most important or desirable of roles to be played. High grades and doctorates were no longer enough to guarantee success outside an academic world itself poised on the brink of redundancy, and corporate executives were increasingly small pawns in global economic games. As for the attractions of public service, who was going to disagree with the young women who said that "politics is all talk and no action"? They simply felt they had better things to do.

Some of these things were far more lucrative as well: in the twenty years after 1970, the number of women–owned small businesses went from 5 percent to 32 percent in the United States, and in Britain nearly 25 percent of the self–employed were women by 1994, twice as many as in 1980. Taking the skills, contacts, and experience gained in their periods of paid employment, these women have tended to be far more successful than their self-employed male counterparts: in the United States, where most new businesses failed, those which were owned by women enjoyed an 80 percent success rate and employed more people than the companies on the *Fortune* 500 list.

Having had little option but to continually explore new avenues, take risks, change jobs, learn new skills, work independently, and drop in and out of the labor market more frequently than their male colleagues, women seem far "better prepared, culturally and psychologically" for the new economic conditions which have emerged at the end of the twentieth century. They are advanced players of an economic game for which self-employment, part-time, discontinuous work, multiskilling, flexibility, and maximal adaptability were suddenly crucial to survival. Women had been ahead of the race for all their working lives, poised to meet these changes long before they arrived, as though they always had been working in a future which their male counterparts had only just begun to glimpse. Perhaps they really were the second sex, if seconds come after firsts.

" 'Let the man get some sleep, Armitage,' Molly said from her futon, the components of the fletcher spread on the silk like some expensive puzzle. 'He's coming apart at the seams.' "

William Gibson, *Neuromancer*

4
3

But there was much more to come. Abandoned by the economic power and social privilege which once made them such attractive, even necessary, mates, the sperm counts fell, birth rates crashed, and the hormonal energy and muscular strength which once served them so well were now becoming liabilities. Women were becoming mothers on their own terms, or not at all. Heterosexual relations were losing their viability, queer connections were flourishing, the carnival had begun for a vast range of paraphilias and so-called perversions, and if there was more than one sex to have, there were also more than two to be. Anything claiming to be normal had become peculiar.

"He was thoroughly lost now; spatial disorientation held a peculiar horror for cowboys."

William Gibson, *Neuromancer*

It was falling apart. They were coming undone. Everything was moving much too fast. What had once seemed destined to become a smoothly regulated world was suddenly running away with itself. Control was slipping through the fingers of those who had thought it was in their hands. Something was wrong. They were losing it all: their senses of security and identity, their grip, the plot, and even their jobs. Couldn't see the point to anything. What else could the masters of the old white world do but redouble their efforts, intensify their drives for security, heighten and perfect their powers? But the more they struggled to adapt and survive, the faster the climate seemed to change. The more they tried to regain control, the more their narrative lost its thread; the closer they came to living the dream, the weaker their grasp on power became. Was it even possible that, regardless of their labors, their hopes and dreams, they had been "the sex organs of the machine world, as the bee of the plant world, enabling it to fecundate and to evolve ever new forms"? All that time, the effort and the pain, the trouble they had taken to maintain control.

"And instead they watch the machines multiply that push them little by little beyond the limits of their nature. And they are sent back to their mountain tops, while the machines progressively populate the earth. Soon engendering man as their epiphenomenon."

Luce Irigaray, *Marine Lover*

cultures

Nothing takes the credit—or the blame—for either the runaway tendencies at work or the attempts to regulate them. Political struggles and ideologies have not been incidental to these shifts, but cultures and the changes they undergo are far too complex to be attributed to attempts to make them happen or hold them back. This is not because some other determination has come into play. If anything does emerge from the complexity of current shifts, it is the realization that cultures cannot be shaped or determined by any single hand or determining factor. Even conceptions of change have changed. Revolution has been revolutionized. There is no center of operations, no organizing core; there are no defining causes, overriding reasons, fundamental bases, no starting points or prime movers; no easy explanations, straightforward narratives, simple accounts, or balanced books. Any attempt to deal with some particular development immediately opens onto them all.

The impossibility of getting a grip, and grasping the changes underway is itself one of the most disturbing effects to emerge from the current mood of cultural change. The prospect of being in a position to know, and preferably control, changes manifest on the social scale has been crucial to modern conceptions of what used to be called man's place in the grand scheme of things. Technology itself was supposed to be a vital means of exerting this explanatory and organizational power. But the revolutions in telecommunications, media, intelligence gathering, and information processing they unleashed have coincided with an unprecedented sense of disorder and unease,

not only in societies, states, economies, families, sexes, but also
in species, bodies, brains, weather patterns, ecological systems.
There is turbulence at so many scales that reality itself seems
suddenly on edge. Centers are subsumed by peripheries, main-
streams overwhelmed by their backwaters, cores eroded by the
skins which were once supposed to be protecting them. Organizers have found themselves eaten up by whatever they were
trying to organize. Master copies lose their mastery, and everything valued for its size and strength finds itself overrun by
microprocessings once supposed too small and insignificant to
count.

nets

Of all the media and machines to have emerged in the late
twentieth century, the Net has been taken to epitomize the
shape of this new distributed nonlinear world. With no limit to
the number of names which can be used, one individual can
become a population explosion on the Net: many sexes, many
species. Back on paper, there's no limit to the games which can
be played in cyberspace. Access to a terminal is also access to
resources which were once restricted to those with the right
face, accent, race, sex, none of which now need be declared.
Using the Net quickly became a matter of surfing, a channel-
hopping mode facilitated and demanded by information which
is no longer bound together in linear texts or library classifications, but instead needs to be laterally traversed.

As the system began to spill out into wider academic usage
over the course of the next twenty years, other networks also
emerged. Businesses developed local, and then wide area net-

works; commercial on-line services appeared; electronic mail and bulletin boards proliferated alongside fanzines and the samizdat press. While the network was doubling in size every year, the screens were gray, the options limited, and the number of users relatively small until the late 1980s. Access was hardly limited to students, hackers, and academics, but certain skills and commitments to computing were prerequisites of any tangible input into the system, and the users of the network occupied a strange frontline between state institutions and anarchic private use. In the wake of a massive expansion of the Net, the arrival of cybercafes, public terminals, falling costs, and a complex of other economic and cultural tendencies, use of the Net has grown not only in the West but in almost two hundred countries of the world. Usenet gives readers and writers access to thousands of articles in thousands of threads in vast populations of newsgroup conversations, continually adding to themselves and fading out of use. On-line worlds scrolled down the screens in IRC (Internet Relay Chat) networks, MUDs (Multi-User Dungeons, or Domains), and MOOs (MUDs Object Oriented), where softbots—software robots—and pseudonymous users interact in labyrinthine virtual worlds. With the development of the World Wide Web, a user-friendly, interactive, multimedia interface which uses Hypertext Markup Language (HTML) to map and interlink the information on the screen to another, and in principle, *any* other site, the Net gained both a gleaming corporate mall, and also a degree of interconnectivity which has continually drawn more computers, pages, links, users, and characters into a network which soon hosted galleries, libraries, shopping malls, company showcases, S&M dungeons, university departments, personal diaries, fanzines . . . every page linked to at least one other, sometimes hundreds, and always proliferating.

The Net has not caught up with the more expansive hopes of unfettered, free-flowing information which were once attached to it. But the technical potential it opens up comes close to the enormous system of lateral cross-referencing which the hypertext networks Ted Nelson first named *Xanadu* in the 1960s, and the system Vannevar Bush called the *memex* in the 1940s. Both these conceptions were far more interactive than the system of the mid-1990s allows. The user of Bush's imagined system left "a trail . . . of interest through the maze of materials available," adding links and connections, inserting passages, and making routes through an immense virtual library whose composition continually shifts as a consequence of the activity of those who are using it. Ted Nelson's envisaged system, which, to some extent, has been realized by the World Wide Web, has the enormous advantage of facilitating this same level of influence with the introduction of (very) small payments of electronic cash for the use of material on specific sites. With the flat-rate subscription system currently in place, links have to be deliberately made and do not, as with pathways across a field of grass, emerge from the sheer force of numbers making them.

As well as potentially facilitating new modes of information circulation, this grass-roots commerce poses great threats to the corporate interests currently in play. But if large-scale commercial activity tends to turn the Net into a shopping mall, it had its beginnings in 1969 as ARPAnet, a U.S. military defense project which quickly joined cockroaches on the short list of those most likely to survive nuclear attack. Developed at the height of the cold war, the Net had also learned from the Viet Cong, whose networks of tunnels and guerrilla techniques had forced a centralized U.S. military machine to adopt unprecedented tactics of distribution and dispersal in response. These military influences on the Net are betrayed in its messages' ability to route and reroute themselves, hunting for ways round obstacles, seeking out shortcuts and hidden passages, continually requisitioning supplies and hitching as many rides as possible. The network and its traffic are so dispersed that any damage to one part of the system, or even a particular message, will have little effect on the whole machinery. Information is transmitted in packets which rarely take the same route twice, and may take many different routes to a destination at which they weave themselves together again. Maps of the network cannot be stolen, not because they are closely guarded, but because there is no definitive terrain. Any map adds to the network and is always already obsolete.

The growth of the Net has been continuous with the way it works. No central hub or command structure has constructed it, and its emergence has been that of a parasite, rather than an organizing host. It has installed none of the hardware on which it works, simply hitching a largely free ride on existing computers, networks, switching systems, telephone lines. This was one of the first systems to present itself as a multiplicitous, bottom-up, piecemeal, self-organizing network which, apart from a quotient of military influence, government censorship, and corporate power, could be seen to be emerging without any centralized control. Not that such lateral networks or bootstrapped systems have "an irresistible revolutionary calling . . ." The leading corporations are now expending all their energies on processes of molecularization and virtualization, continually downsizing and turning themselves into flattened horizontal operations and, in effect, getting all such modes of activity on their side. No matter how spontaneous their emergence, self-organizing systems are back in organizational mode as soon as they have organized themselves.

This conflict is inscribed in the double-edged quality of

the word itself. Technology is both a question of logic, the long arm of the law, *logos*, "the faculty which distinguishes parts ('on the one hand and on the other hand')," and also a matter of the skills, digits, speeds, and rhythms of techno, engineerings which run with "a completely other distribution which must be called nomadic, a nomad *nomos*, without property, enclosure or measure." The same ambivalence is inscribed in the zeros and ones of computer code. These bits of code are themselves derived from two entirely different sources, and terms: the binary and the digital, or the symbols of a logical identity which does indeed put everything on one hand or the other, and the digits of mathematics, full of intensive potential, which are not counted by hand but on the fingers and, sure enough, arrange themselves in pieces of eight rather than binary pairs.

5
0

The techno and the digital are never perceived to run free of the coordinating eyes and hands of logic and its binary codes. But logic is nothing without their virtual plane. They are the infrastructure to its superstructure: not another order of things, but another mode of operations altogether, the matters of a distribution which is "demonic rather than divine, since it is a peculiarity of demons to operate in the intervals between the gods' fields of action . . . thereby confounding the boundaries between properties."

"You know I am a d——d ODD animal! And as my mother often says, she never has quite yet made up her mind if it is the Devil or Angel that watches peculiarly over me; only that it IS one or the other, without doubt!

"(And for my part, I am quite indifferent which.)"

Ada Lovelace, December 1841

digits

The vast majority of what are now assumed to be the West's mathematical terms and axioms are either Arabic or Hindu. The word algebra is taken from the title of the *Al-gebr we'l mukabala*, a book written in the ninth century by one of the most sophisticated Arab mathematicians, Alkarismi, who gave his name to the algorithm. The *Al-gebr* is in turn based on the work of Brahmagupta, a Hindu mathematician and astronomer who, in the seventh century, consolidated India's sophisticated but unwieldy arithmetical principles in the form of twenty basic processes "essential to all who wish to be calculators."

The system of notation and calculation which emerged from this fusion of Hindu and Arabic arithmetic was introduced to the West by both Arabic scholars and Asian traders. Indian arithmetic had already been carried by merchants as far west as Baghdad, and Alkarismi's own arithmetical prowess is said to have resulted from his own travels in India. It was a great space-saving device when compared to its far more cumbersome counterparts, most of which had been developed in conjunction with the abacus, a device which was unknown to Hindu culture, but had been widely used in the Egyptian, Babylonian, Greek, and Roman worlds. While the abacus had removed the need to process and store numbers in concise written form, India had developed a sophisticated system of notation which it used both to calculate and record results.

India had effectively developed a written abacus, using its written numbers in place of pebbles or beads, giving them the same signs regardless of the positions they assumed, and using 0 or a dot to indicate an empty column of the virtual abacus. Whereas abacists used completely different signs for numbers with different place values—such as I for one and X for ten in Roman numerals—the Hindu system could use the same digit—1—to compose one, ten, hundred, and an obviously vast number of other numbers.

"It is India that gave us the ingenious method of expressing all numbers by means of ten symbols," wrote Pierre-Simon Laplace, "each symbol receiving a value of position as well as an absolute value." In other words, the numbers were both cardinal and ordinal, each expressing its place in the string (first, second, third etc.), as well as a value specific to itself. Unlike the Roman numerals, in which two is simply two ones collected together, the Sanskrit two is a qualitatively different number to one, an entity or character in its own right. As Laplace points out, the new arithmetic was "a profound and important idea which appears so simple to us now that we ignore its true merit, but its very simplicity, the great ease which it has lent to all computations, puts our arithmetic in the front rank of useful inventions." Although this statement about "our arithmetic" subtly appropriates the new system as one of the West's "inventions," Laplace continues, "We shall appreciate the grandeur of this achievement when we remember that it escaped the genius of Archimedes and Apollonius, two of the greatest men produced by antiquity."

"Certainly my troops must consist of numbers, or they can have no existence at all, & would cease to be the particular sort of troops in question.—But then what are these numbers? There is a riddle."

Ada Lovelace

To a Europe still counting in bundles of Roman sticks, this new arithmetic, with its alien Sanskrit figures, was an infidel system which posed an extraordinary threat to the stability of the Western world. Although the Eastern system is as widely used as the alphabet today, it was not until the Renaissance that Europe's new merchants overcame the opposition of the Church to the introduction of the numbers 123456789 and 0. One of the first texts on the new arithmetic—which was also one of the first books in the English language, *The Craft of Nombrynge (ca. 1300)*—was composed while edicts forbidding the use of the numbers were still being issued in Florence. By 1478, the first manual on the new arithmetic had been printed in Italy on one of the then brand-new Gutenberg presses. "Numeration is the representation of numbers by figures," it explained. "This is done by means of ten letters or figures as here shown, 1.,2.,3.,4.,5.,6.,7.,8.,9.,0. Of these the first figure, 1, is not called a number, but the source of number. The tenth figure, 0, is called cipher or 'nulla,' i.e. the figure of nothing, since by itself it has no value, although when joined with others it increases their value."

In addition to its numbers, the new arithmetic introduced negative numbers and irrational numbers, as well as zero and the decimal point. These were features crucial to the networks of banking and trade which became increasingly important to European culture in the fifteenth century. Trade, which is now widely assumed to be a peculiarly Western invention, was then as new to Europe as these numbers, and there is little doubt that even the simple matters of keeping accounts, setting prices, doing deals, and working with large numbers were simply impossible with Roman numerals. This was not the least of the reasons why the infidel arithmetic threatened a Christian culture which, even now, demurs at the thought of Sunday trade.

The one of the new arithmetic was also very different to the old straight line which had figured as both a number and the ninth letter of the Roman alphabet. Western philosophy is supposed to be an elucidation and confirmation of the unity of one, a number which had been held in great esteem long before there was one male god. To the ancient Greeks, one was everything and anything, first and last, best and good, universal, unified. It was the symbol of existence, identity, and being. Strictly speaking, there was nothing else. To be anything at all was to be one.

For all its dreams of self-sufficiency, even one has always needed another of some kind. But since it was the only one as well, it had to ensure that any other options were merely impoverished variations on its theme. The Greeks recognized many as an alternative to one but, like the Romans, even this was conceived as a collection of many ones. Derived from the Greek term *iota,* and closely related to atom and jot, this one was taken to symbolize any individuated and indivisible entity, whereas the Sanskrit one functioned in relation to the other eight digits of the Hindu system. But one closely resembled the old Roman line and was easily subsumed into the old paradigm. Any differences between the two were more or less erased.

Zero posed a very different threat. When it first appeared in the new string of infidel figures, the old Church fathers did everything to keep it out of a world which then revolved around one and its multiples: one God, one truth, one way, one one. The numbers 2,3,4,5,6,7,8,9 were subversive enough, but zero was unthinkable. If it wasn't one of something, it couldn't be allowed. Then again, the Church could hardly be seen to protest too much about something that, as far as they could see, wasn't really there at all. If zero was nothing, it should be as easy to absorb as the Sanskrit one had been. And, sure enough, zero was appropriated as a sign of absence, nonbeing, and nothingness. The ancient unity of something and nothing was apparently undisturbed.

holes

"Somewhere there is a siren. Her green body is covered with scales. Her face is bare. The undersides of her arms are a rosy colour. Sometimes she begins to sing. The women say that of her song nothing is to be heard but a continuous 0. That is why this song evokes for them, like everything that recalls the 0, the zero or the circle, the vulval ring."

Monique Wittig, *Les Guérillères*

Having escaped the rigors of an education which would have taught her not to ask such things, Ada wandered off, around and about, and wondered about zero too. One of her earliest enquiries to Augustus De Morgan, her tutor in mathematics, concerned the status of this figure. Did it exist as a "thing," she asked? Was it something, or nothing, or something else again? He gave her an intriguing answer. "Zero is *something*," he explained, "though not some *quantity*, which is what you here mean by *thing*."

"She does not set herself up as one, as a (single) female unit. She is not closed up or around one single truth or essence. The essence of a truth remains foreign to her. She neither has nor is a being. And she does not oppose a feminine truth to a

masculine truth . . . the female sex takes place by embrac-
ing itself, by endlessly sharing and exchanging its lips, its
edges, its borders, and their 'content,' as it ceaselessly be-
comes other, no stability of essence is proper to her."
Luce Irigaray, *Speculum of the Other Woman*

Zero may mean nothing to the Western world, but this has
nothing to do with the way it works. It was certainly crucial to
the functioning of the Analytical Engine, a machine which,
according to Menabrea, used an "occult principle of change"
which allowed it to "provide for singular values." The Engine
was able to deal with those functions "which necessarily change
in nature when they pass through zero or infinity, or whose
values cannot be admitted when they pass these limits. When
such cases present themselves, the machine is able, by means of a
bell, to give notice that the passage through zero or infinity is
taking place, and it then stops until the attendant has again set it
in action for whatever process it may next be desired to perform. If this process has been foreseen, then the machine instead
of ringing, will so dispose itself as to present the new cards
which have relation to the operation that is to succeed the
passage through zero and infinity." It is the possibility of this
passage which allows the machine "arbitrarily to change its
processes at any moment, on the occurrence of any specified
contingency."

In terms of the pragmatic roles they play, the zeros and
ones of machine code do far more than hark back to the binaries their logical symbols represent. If zero is supposed to signify
a hole, a space, or a missing piece, and one is the sign of
positivity, digital machines turn these binaries around. In both
electronic systems and the punched cards of weaving machines,
a hole is one, and a blank is zero, in which case there are two

missing elements, if missing is where either can be said to go. No longer a world of ones and not-ones, or something and nothing, thing and gap, but rather not-holes and holes, not-nothing and nothing, gap and not-gap. Not that this matters any more than the initial dualism between one and a zero conceived as not-one. Zero was always something very different from the sign which has emerged from the West's inability to deal with anything which, like zero, is neither something in particular nor nothing at all. And it is certainly the case that, with or without the signs that represent them as inert negativities, holes themselves are never simply absences of positive things. This is a purely psychoanalytical myth. For Deleuze and Guattari, it is not even enough "to say that intense and moving particles pass through holes; a hole is just as much a particle as what passes through it . . ." Holes are not absences, spaces where there should be something else. "Flying anuses, speeding vaginas, there is no castration." Adrift in the doped lattices of a silicon crystal, a hole is a positive particle before it is the absence of a negatively charged electron, and the movement of electrons toward the positive terminal is also a flow of holes streaming back the other way. Holes are charged particles running in reverse. For the quantum physicist, "holes are not the absence of particles but particles traveling faster than the speed of light."

"Transpierce the mountains instead of scaling them, excavate the land instead of straiting it, bore holes in space instead of keeping it smooth, turn the earth into swiss cheese."

Gilles Deleuze and Félix Guattari, *A Thousand Plateaus*

cyborg manifestos

For years, decades, centuries, it seemed as though women were lagging behind the front runners of the human race, struggling to win the rights attained by men, suffering for want of the status which full membership of the species would supposedly have given them. And as long as human was the only thing to be, women have had little option but to pursue the possibility of gaining full membership of the species "with a view to winning back their own organism, their own history, their own subjectivity." But this is a strategy which "does not function without drying up a spring or stopping a flow." And there are processes of parallel emergence, noncausal connections and simultaneous developments which suggest that sexual relations continually shift in sympathy with changes to the ways many other aspects of the world work. If Simone de Beauvoir's *Second Sex* found itself compelled to call for "men and women" to "univocally affirm their brotherhood" in 1949, this was also the point at which the first sex began to find itself subsumed by self-organizing tendencies beyond its ken or its control. By 1969, when Monique Wittig published *Les Guérillères,* these tendencies were emerging as networks which didn't even try to live up to the existing definitions of what it was to be a proper one of anything at all. And by the 1970s, when Luce Irigaray wrote *This Sex Which Is Not One,* fluid complexities were giving a world which had once revolved around ones and others a dynamic which obsolesced the possibility of being one of anything at all.

As personal computers, samplers, and cyberpunk narratives proliferated in the mid-1980s, Donna Haraway's cyborgs were writing manifestos of their own. "By the late twentieth century," they declared, "our time, a mythic time, we are all chimeras, theorized and fabricated hybrids of machine and organism; in short, we are all cyborgs." And while the shiny screens of the late twentieth century continued to present themselves as clean-living products of the straight white lines of a peculiarly man-made world, Haraway's text excited a wave of subversive female enthusiasm for the new networks and machines. In the early 1990s, a cyberfeminist manifesto appeared on an Australian billboard and declared, "The clitoris is a direct line to the matrix," a line which refers to both the womb—*matrix* is the Latin term, just as *hystera* is the Greek—and the abstract networks of communication which were increasingly assembling themselves.

"You may not encounter ALL NEW GEN as she has many guises. But, do not fear, she is always in the matrix, an omnipresent intelligence, anarcho cyber terrorist acting as a virus of the new world disorder."

VNS Matrix

They say she wears "different veils according to the historic period." They say her "original attributes and epithets were so numerous . . . in the hieroglyphics she is called 'the many-named,' 'the thousand-named' . . . 'the myriad-named.'" They say, "the future is unmanned." They say, "let those who call for a new language first learn violence. They say, let those who want to change the world first seize all the rifles. They say that they are starting from zero. They say that a new world is beginning." They say, "if machines, even the machines of theory, can arouse themselves, why not women?"

programming language

"It is already getting around—at what rate? in what contexts? in spite of what resistances?—that women diffuse themselves according to modalities scarcely compatible with the framework of the ruling symbolics. Which doesn't happen without causing some turbulence, we might even say whirlwinds, that ought to be reconfined within solid walls of principle, to keep them from spreading to infinity . . ."

Luce Irigaray, *This Sex Which Is Not One*

In May 1979, Commander John D. Cooper came up with a name which the United States Department of Defense's High Order Language Working Group (HOLWG) could accept for their new programming language: Ada, chosen "in honor of an obscure but talented mathematician, Ada, Countess of Lovelace." When HOLWG approached the Earl of Lytton, one of Ada's descendants, for permission to use the name, he "was immediately enthsiastic about the idea and pointed out that the letters 'Ada' stood 'right in the middle of "radar." ' "

shuttle systems

There is always a point at which, as Freud admits, "our material—for some incomprehensible reason—becomes far more obscure and full of gaps." And, as it happens, Freud's weaving women had made rather more than a small and debatable contribution to his great narrative of inventions and discoveries. Far more than a big and certain one as well. It is their micro-processes which underlie it all: the spindle and the wheel used in spinning yarn are the basis of all later axles, wheels, and rotations; the interlaced threads of the loom compose the most abstract processes of fabrication. Textiles themselves are very literally the softwares linings of all technology.

String, which has been dated to 20,000 B.C., is thought to be the earliest manufactured thread and crucial to "taking the world to human will and ingenuity," not least because it is such multipurpose material. It can be used for carrying, holding, tying, and trapping, and has even been described as "the unseen weapon that allowed the human race to conquer the earth." Textiles underlie the great canvases of Western art, and even the materials of writing. Paper now tends to be made from wood, but it too was woven in its early form, produced from the dense interlacing of natural fibers. The Chinese, with whom the production of paper is thought to have begun some 2,000 years ago, used bamboo, rags, and old fishing nets as their basic materials; papyrus, from which the word paper is itself derived, was used in ancient Egypt, and later Arab cultures used the same flax from which linen is produced. Wood pulp gradually took over from the rags which Europe used until the nineteenth century, and most paper is now produced from fibers which are pulped and bleached, washed and dried, and then filtered onto a mesh and compressed into a fine felt.

Evidence of sophisticated textile production dates to 6,000 B.C. in the southeast regions of Europe, and in Hungary there is evidence that warp-weighted looms were producing designs of extraordinary extravagance from at least 5,000 B.C. Archaeological investigations suggest that from at least the fourth millennium B.C. Egyptian women were weaving linen on horizontal looms, sometimes with some two hundred threads per inch, and capable of producing cloths as wide as nine feet and seventy-five feet long. Circular warps, facilitating the production of seamless tubes for clothing, and tapestry looms, able to weave the dense complications of images visible in weft threads so closely woven as to completely conceal the warps, were also in use in ancient Egypt where, long before individual artisans stamped their work with their own signatures, trademarks and logos were woven in to indicate the workshop in which cloths had been produced. Cloths were used as early currency, and fine linens were as valuable as precious metals and stones. In China, where the spinning wheel is thought to have first turned, sophisticated drawlooms had woven designs which used thousands of different warps at least two and a half thousand years before such machines were developed in the West.

It may be a bare necessity of life, but textiles work always goes far beyond the clothing and shelter of the family. In terms of quality, sophistication, and sheer quantity, the production of textiles always seems to put some kind of surplus in play. The production of "homespun" yarn and cloth was one of the first cottage industries, pin money was women's earliest source of independent cash, and women were selling surplus yarn and cloth and working as small-scale entrepreneurs long before the emergence of factories, organized patterns of trade, and any of the mechanisms which now define the textiles industry. Even when cloths and clothes can be bought off the rack, women continue to absorb themselves in fibrous fabrications.

There is an obsessive, addictive quality to the spinning of yarn and the weaving of cloth; a temptation to get fixated and locked in to processes which run away with themselves and those drawn into them. Even in cultures assumed to be subsistence economies, women who did only as much cooking,

cleaning, and childcare as was necessary tended to go into overdrive when it came to spinning and weaving cloth, producing
far more than was required to clothe and furnish the family
home. With time and raw materials on their hands, even "Neolithic women were investing large amounts of extra time into
their textile work, far beyond pure utility," suggesting that not
everything was hand to mouth. These prehistoric weavers seem
to have produced cloths of extraordinary complexity, woven
with ornate designs far in excess of the brute demand for simple
cloth. And wherever this tendency to elaboration emerged, it
fed into a continual exploration of new techniques of dyeing,
color combination, combing, spinning, and all the complications of weaving itself.

Even in Europe there had been several early and sophisticated innovations. Drawlooms had been developed in the Middle Ages, and while many of Leonardo da Vinci's "machines for
spinning, weaving, twisting hemp, trimming felt, and making
needles" were never made, he certainly introduced the flyer and
bobbin which brought tension control to the spinning wheel.
Unlike "the spinster using the older wheel," she now "slack-
ened her hold on the yarn to allow it to be wound on to the
bobbin as it was being twisted."

It is often said that Leonardo's sixteenth-century work
anticipated the industrial revolution "in the sense that his 'machines' (including tools, musical instruments, and weapons) all
aspired toward systemic automation." But it was his intuition
that textiles machines were "more useful, profitable, and perfect
than the printing press" which really placed him ahead of his
time. If printing had spread across the modern world, textiles
led the frantic industrialization of the late eighteenth and early
nineteenth centuries. "Like the most humble cultural assets,
textiles incessantly moved about, took root in new re-

gions . . ." The first manufactory was a silk mill on an island in the Derwent near Derby built early in a century which also saw the introduction of the spinning jenny, the water frame, the spinning mule, the flying shuttle, the witches' loom, and the power loom. A spiral of "inventions in both spinning and weaving (interacting and mutually stimulating) had attracted capital, concentrated labour, increased output and swollen imports and exports." This was cloth capitalism, a runaway process which quite literally changed the world. In the 1850s, it was said that "if Providence had never planted the cotton shrub those majestic masses of men which stretch, like a living zone, through our central districts, would have felt no existence; and the magic impulse which has been felt . . . in every department of national energy, our literature, our laws, our social condition, our political institutions, making us almost a new people, would never have been communicated." Textiles had not merely changed the world: they seemed to have mutated its occupants as well. *Almost a new people . . ."* "I was surprised at the place but more so at the people," wrote one commentator of Birmingham, the site of the first cotton-spinning mill. "They were a species I had never seen."

While the industrial revolution is supposed to have made the break between handheld tools and supervised machines, the handmade and the mass-produced, the introduction of technology to more primitive textiles techniques is both a break with the old ways and a continuation of the lines on which the women were already at work. Even before its mechanization, the loom was described as the "most complex human engine of them all," not least because of the extent to which it "reduced everything to simple actions: the alternate movement of the feet worked the pedals, raising half the threads of the warp and then the other, while the hands threw the shuttle carrying the thread of the woof." When John Heathcote, who patented a lace-making machine just after Jacquard built his loom, first saw "a woman working on a pillow, with so many bobbins that it seemed altogether a maze," his impression was that lace was a "heap of chaotic material." In an attempt to unravel the mystery, he "drew a thread, which happened to draw for an inch or two longitudinally straight, then started off diagonally. The next drew out straight. Then others drew out in various directions. Out of four threads concurring to make a mesh, two passed one way, the third another and the fourth another still. But at length I found they were in fact used in an orderly manner . . ." It was then a matter of producing "a fabric which was an exact imitation of the thread movements of handmade lace." This is both the ordering of chaos, and also how its networks replicate themselves.

There were other spin-offs from textiles too. The weaving of complex designs demands far more than one pair of hands, and textiles production tends to be communal, sociable work allowing plenty of occasion for gossip and chat. Weaving was already multimedia: singing, chanting, telling stories, dancing, and playing games as they work, spinsters, weavers, and needle-workers were literally networkers as well. It seems that "the women of prehistoric Europe gathered at one another's houses to spin, sew, weave, and have fellowship." Spinning yarns, fabricating fictions, fashioning fashions . . . : the textures of woven cloth functioned as means of communication and information storage long before anything was written down. "How do we know this? From the cloth itself." This is not only because, like writing and other visual arts, weaving is often "used to mark or announce information" and "a mnemonic device to record events and other data." Textiles do communicate in terms of the images which appear on the right side of the cloth, but this is only the most superficial sense in which they process and store data. Because there is no difference between the process of weaving and the woven design, cloths persist as records of the processes which fed into their production: how many women worked on them, the techniques they used, the skills they employed. The visible pattern is integral to the process which produced it; the program and the pattern are continuous.

Information can be stored in cloth by means of the meaningful messages and images which are later produced by the pen and the paintbrush, but data can also be woven in far more pragmatic and immediate ways. A piece of work so absorbing as a cloth is saturated with the thoughts of the people who produced it, each of whom can flash straight back to whatever they were thinking as they worked. Like Proust's madeleines, it carries memories of an intensity which completely escapes the written word. Cloths were also woven "to 'invoke magic'—to protect, to secure fertility and riches, to divine the future, perhaps even to curse," and in this sense the weaving of spells is far more than a metaphorical device. "The weaver chose warp threads of red wool for her work, 24 spun one direction, 24 spun the other way. She divided the bunch spun one way into 3 sets of 8, and the other bunch into 4 sets of 6, and alternated them. All this is perhaps perfectly innocent, but . . ."

If the weaving of such magical spells gives priority to the process over the completion of a task, this tendency is implicit in the production of all textiles. Stripes and checks are among the most basic of colored and textured designs which can be woven in. Both are implicit in the grids of the woven cloth itself. Slightly more complex, but equally integral to the basic web, are the lozenges, or diamonds, still common in weaves across the world. These open diamonds are said to indicate fertility and tend to decorate the aprons, skirts, and belts which

are themselves supposed to be the earliest forms of clothing.
"These lozenges, usually with little curly hooks around the
edge, rather graphically, if schematically, represent a woman's
vulva." These images are quite unlike those which are later
painted on the canvas or written on the page. The lozenge is
emergent from the cloth, diagonal lines implicit in the grids of
the weave. And even the most ornate and complex of woven
designs retains this connection to the warps and wefts. When
images are later painted, or written in the form of words on a
page, patterns are imposed on the passive backdrop provided by
the canvas or the page. But textile images are never imposed on
the surface of the cloth: their patterns are always emergent from
an active matrix, implicit in a web which makes them immanent to the processes from which they emerge.

As the frantic activities of generations of spinsters and
weaving women makes abundantly clear, nothing stops when a
particular piece of work has been finished off. Even when magical connections are not explicitly invoked, the finished cloth,
unlike the finished painting or the text, is almost incidental in
relation to the processes of its production. The only incentive to
cast off seems to be the chance completion provides to start
again, throw another shuttle, cast another spell.

As writing and other visual arts became the privileged
bearers of memory and messages, weaving withdrew into its
own screens. Both canvases and paper reduce the complexities
of weaving to raw materials on which images and signs are
imposed: the cloths from which woven patterns once emerged
now become backcloths, passive matrices on which images are
imposed and interpreted as if from on high. Images are no
longer carried in the weave, but imprinted on its surface by the
pens and brushes with which shuttles become superficial carriers of threads. Guided by the hand—eye coordinations of what

are now their male creators, patterns become as individuated and unique as their artists and authors. And whereas the weave was once both the process and the product, the woven stuff, images are now separated out from matrices to which they had been immanent. The artist sees only the surface of a web which is covered as he works; the paper on which authors now look down has no say in the writing it supports.

The processes themselves become dematerialized as myths, legends, and metaphors. Ariadne's thread, and the famous contest in which the divine Athena tore mortal Arachne's weaving into shreds, are among the many mythical associations between women and webs, spinsters and spiders, spinning yarns and storylines. For the Greeks, the Fates, the Moirai, were three spinsters—Klotho, Lachesis, and Atropos—who produced, allotted, and broke the delicate contingency of the thread of life. In the folktales of Europe, spindles become magic wands, Fates become fairies, and women are abandoned or rescued from impossible spinning and weaving tasks by supernatural entities, godmothers and crones who transform piles of flax into fine linen by means more magical than weaving itself, as in "Rumpelstiltskin," "The Three Spinsters," and "The Sleeping Beauty." "European folktales are full of references to the making of magical garments, especially girdles, in which the magic seems to be inherent in the weaving, not merely in special decoration."

As for the fabrics which persist: evaluated in these visual terms, their checks and diagonals, diamonds and stripes become insignificant matters of repeating detail. This is why Freud had gazed at work which was so literally imperceptible to him. Struggling only to interpret the surface effects of Anna's work as though he was looking at a painting or a text, the process of weaving eluded him: out of sight, out of mind, out of his world.

This was a process of disarmament which automation should have made complete. But if textiles appear to lose touch with their weaving spells and spans of time, they also continue to fabricate the very screens with which they are concealed. And because these are processes, they keep processing. "Behind the screen of representation," weaving wends its way through even the media which supplant it. While paper has lost its associations with the woven fabrics with which it began, there are remnants of weaving in all writing: yarns continue to be spun, texts are still abbreviated textiles, and even grammar—glamor—and spelling retain an occult connectivity. Silkscreens, printing presses, stencils, photographic processes, and typewriters: by the end of the nineteenth century images, texts, and patterns of all kinds were being processed by machines which still used matrices as means to their ends, but also repeated the repeating patterns downgraded by the one-off work of art. And while all these modes of printing were taking technologies of representation to new heights, they were also moving on to the matrices of times in which these imprinting procedures would reconnect with the tactile depth of woven cloth.

casting on

Spinning is "a perilous craft" wrote Mircea Eliade. "The moon 'spins' Time and 'weaves' human lives. The Goddesses of Destiny are spinners." When he looks at the seclusion of pubescent girls and menstruating women, often the occasion for the spinning of both actual and fictional yarns, he detects "an occult connection between the conception of the periodical creations of the world . . . and the ideas of Time and Destiny, on the one hand, and on the other, nocturnal work, women's work, which has to be performed far from the light of the sun and almost in secret. In some cultures, after the seclusion of the girls is ended they continue to meet in some old woman's house to spin together." And wherever spinning is ubiquitous, there is often "a permanent tension, and even conflict, between the groups of young spinning girls and the men's secret societies. At night the men and their gods attack the spinning girls and destroy not only their work, but also their shuttles and weaving apparatus."

If the psychoanalysts provide the only accounts of hysteria, the only records of the witch-hunting which swept three centuries of premodern society are written by the hunters and from their point of view. "The voices of the accused reach us strangled, altered, distorted; in many cases, they haven't reached us at all." What "really happened" has left the scene. Historians of witchcraft "have implicitly or explicitly derived the subject of their research from the interpretative categories of the demonologists, the judges or witnesses against the accused," and "with very few exceptions," most scholarly studies "have continued to concentrate almost exclusively on persecution, giving little or no attention to the attitudes and behaviour of the persecuted." Even feminist scholars have endorsed this approach. "Clearly," writes Mary Daly of those on trial, "the supposed sexual fantasies of these women were (are) archetypically male fantasies," and the accused were nothing more than "projection screens for these hallucinations."

If everything remaining of the witch cults is circumscribed by those who define and prosecute their crimes, anyone "declining to restrict himself to recording the results of this historical violence can find fragments, relatively immune from distortions, of the culture that the persecution set out to eradicate."

The prosecution evidence is riddled with gaps: there are holes in the stories, twists to the plots. "Hence—for anyone un-resigned to writing history for the nth time from the standpoint of the victors—the importance of the anomalies, the cracks that occasionally (albeit very rarely) appear in the documentation, undermining its coherence."

All God's children could be led astray, and many men met their deaths at the stake. As in the case of hysteria, the witches were not necessarily male. Persecutors testified to "the exis-tence of an actual sect of female and male witches," who "met at night, generally in solitary places, in fields or on mountains. Sometimes, having anointed their bodies, they flew, arriving astride poles and broom sticks; sometimes they arrived on the backs of animals, or transformed into animals themselves . . ." But the *Malleus Maleficarum,* a fifteenth-century witch-hunter's guide, also reported that "a greater number of witches is found in the fragile feminine sex than among men." It argued that women were particularly predisposed to an "addiction to witchcraft" and considered them "to be of a different nature from men," especially "as regards intellect, or the understanding of spiritual things." Women were said to have "weak memo-ries," so that "it is a natural vice in them not to be disciplined, but to follow their own impulses without any sense of what is due; this is her whole study, and all that she keeps in her mem-ory."

The hunters tied themselves in terrible knots in an attempt to prove both that the witches' activities were real enough to merit the prosecutions, and also that they were simply fantasies. "It cannot be admitted as true that certain wicked women, perverted by Satan and seduced by the illusions and phantasms of devils, do actually, as they believe and profess, ride in the night-time on certain beasts with Diana a goddess of the Pagans, or with Herodias and an innumerable multitude of women, and in the untimely silence of night pass over immense tracts of land, and have to obey her in all things as their Mistress, etc." Flight was simply a delusion: the witches didn't really get in touch with the "innumerable multitude of women" they thought they met. They have believed they went hunting with Diana, Artemis, the Amazon queen, but it was all in the mind, it wasn't happening. "Awakening from sleep, she began a long raving story of crossing seas and mountains, and she brought forth false responses. We denied her story, but she insisted upon it." But, on the other hand, it was this tendency to ascribe the witches' activities to "imagination and illusion" which also suggested that "they were really harmless." And according to the *Malleus Maleficarum,* "For this reason many witches remain unpunished, to the great dispraise of the Creator, and to their own most heavy increase."

"Under the repeated play of movement in the fingers a membrane grows between them that seems to join them, then prolong them, until eventually it extends beyond the hand and descends along the arm, it grows, it lengthens, it gives the women a sort of wing on either side of their body. When they resemble giant bats, with transparent wings, one of them comes up and, taking a kind of scissors from her belt, hastily divides the two great flaps of silk. The fingers immediately recommence their movement."

Monique Wittig, *Les Guérillères*

flight

Ada Lovelace loved all forms of communication. She sometimes
wrote several letters each day, and much of her surviving writing survives in this form. "Think what a delight," she wrote in
a letter when she learned that the electrical telegraph was coming to town in 1844. "Wheatstone says that sometimes friends
hold conversations from one terminus to the other; that one can
send for anyone to speak to one . . . Wonderful agent and
invention!"

At the age of twelve she had entertained hopes of "writing
a book of *Flyology* illustrated with plates," and told her mother
she would "be able to fly about with all your letters and messages and shall be able to carry them with much more speed
than the post or any other *terrestrial* contrivances and to make
the thing quite complete a part of the flying accoutrement shall
be a letter bag, a small compass & a map which the two last
articles will enable me to cut across the country by the most
direct road without minding either mountains, hills, valleys,
rivers, lakes &c, &c, &c. My book of Flyology shall contain a list
of the advantages resulting from flying and it shall also contain a
complete explanation of the anatomy of a bird." Ada had plans
to build her wings from paper or silk stiffened with wire, and
also imagined "a thing in the shape of a horse with a
steamengine in the inside so contrived as to move an immense
pair of wings, fixed on the outside of the horse, in such a
manner as to carry it up into the air while a person sits on its
back."

virtual aliens

"They speak together of the threat they have constituted towards authority, they tell how they were burned on pyres to prevent them from assembling in future."

Monique Wittig, *Les Guérillères*

The "overwhelming majority of electronics assembly jobs are occupied by young female workers on relatively low wages. In this respect, there are clear parallels with the situation in the textiles and clothing industries . . ." Most of these women do "assembly, the bonding of hair-thin wires to semiconductor chips, and the associated packaging. Though the work requires good eyesight and dexterity, little training is required . . ." Silicon Valley, Silicon Glen, Bangalore, Jakarta, Seoul, and Taipei provide dispersed networks of what U.S. multinationals call "virtual aliens" to fabricate the wafers, assemble the circuits, set up the keyboards and the screens, make the chips that make the chips that turn the computers on. They work in the global factory of the new transnationals: "On the west coast, Filipinas, Thais, Samoans, Mexicans and Vietnamese have made the electronics assembly line a microcosm of the global production process."

Microprocessing has always been low status, poorly paid, sometimes dangerous. The terms and conditions of life in the factories and offices may be the smallest of improvements on those of compulsory service in the home. To those who already have room of their own, such moves seem paltry when compared to the rhetoric with which rights are declared and equality is sought. But these infiltrations won their spaces too. The work of these virtual aliens is the latest in the long and twisted line of microprocesses which emerge from a tangle of telephone lines, dials, operators, cables, tones, switches, and plugs; the keys, carriages, and cases of typewriters; the punched-card programs of calculators, pianolas, and looms; flying shuttles, spinning wheels. If she hasn't had a hand in anything, her fingerprints are everywhere.

Left and right, base and superstructure, proletariat and bourgeoisie: like every reproductive system, industrial capitalism was itself supposed to function along the clear-cut binary lines. Often to the great detriment of the working class, the antagonism between forces and modes of production has been played out as a personal argument between the men: a matter of political consciousness, a struggle between bosses and workers, firms and unions, states and revolutionary cells. Organized and organizing factions have confronted each other as two sides of a split identity struggling to reconcile itself in some great climactic moment of revolution, and theories, critiques, and statistics have concentrated on male employment and the fate of the male worker who, together with modern capitalism and its critiques, has been largely engaged in matters of hand–eye coordination. Manual work and man's work have been more or less synonymous, both for the workers—hired hands required to work with their hands, hand tools, handles, and other hand-size components—and the bosses—the ones who manage and manipulate the manufactories, and assume it's all in their hands. This is the binary machine again: two hands and two sides of a game which is supposed to be conducted by another single hand: the invisible hand of capital, perfectly integrated with the supervising eye of the state.

Women, either their own or the proletariat's proletarians,

as Engels called them, have been the least of the bourgeoisie's concerns. Immersed in the low-status microprocesses of textiles production, secretarial work, and the production of miniature components, women are supposed to be the most inconspicuous and insignificant of cogs in the wheels of industry. Women have been off the productive map, out of the dialectical loop: no desire, no agency, not even the alienation of the male worker. Kept apart by the demands of home work, housework, and heterosexual monogamy, the women couldn't get together to organize themselves after the fashion of the men. But for all the instabilities and crises it induced, the industrial proletariat was never the only carrier of revolutionary change, if it was ever such a thing at all. Perhaps its campaigns even served to distract bourgeois man from the really dangerous guerrillas in his midst, those apparently inconspicuous, well-behaved little creatures who spent their time making lists, detailing procedures, typing, sorting, coding, folding, switching, transmitting, receiving, wrapping, packaging, licking the envelopes, fingers in the till.

Women, children, and migrant workers have always been poorly paid, last in, and first out, a reservoir of labor which can be brought on stream as required. They are brought into the factories, the mills, and the new bureaucracies only in response to the demands of booming or war economies, and always under the strict supervision of their male superiors. Both the bosses and the male workers ensure that they are kept away from the important jobs. Managers treat them just like the men, only worse: they are paid, but they are paid less; their work is valued but not as highly as that of their male counterparts. As for their coworkers, the line adopted by America's late–nineteenth century tobacco unions has been repeated time and again: "we have combated from its incipiency the movement of the introduction of female labor in any capacity whatever," they declared. "We can not drive the females out of the trade, but we can restrict this daily quota of labor through factory laws."

cocoons

It has long been assumed in the Western world that technologies are basically tools, means to ends decided in advance by those who make them and put them to use. Whatever the particular purposes for which they are designed and employed, the overriding rationale has always been the effort to secure and extend the powers of those whose interests they are supposed to serve. And their interests have in turn been defined as the exercise of control over something variously defined as nature, the natural, the rest of the world. This crude model of the user and the used has legitimized the scientific projects, colonial adventures, sexual relations, and even the artistic endeavors of the modern world. It continues to inform the deployment of even the most complex machines.

But both man and his tools exist "only in relation to the interminglings they make possible or that make them possible." The user and the used are merely the perceptible elements, the identifiable components which are thrown up by—and serve also to contain—far more complex processes. The weaver and the loom, the surfer and the Net: none of them are anything without the engineerings which they both capture and perpetuate.

These are processes which mock all grandiose attempts to name names and identify great moments of invention and dis-

covery. It is, as Braudel points out, "patient and monotonous
efforts" which lead machinery on. Technical development is
not only a matter of "the brisk changes we are a little too quick
to label revolutions," he writes, "but also the slow improvements in processes and tools . . . those innumberable actions
which certainly have no innovating significance but which are
the fruit of accumulating knowledge: the sailor fixing his ropes,
the miner digging his gallery, the peasant behind his plough, the
smith at his anvil." These are the artisans, technicians, engineers
whose work is more akin to "a collection of recipes drawn from
craftsmen's experience" than a tale of steady progress to some
well-established end, and has "somehow or other evolved un-
hurriedly" by means of its own peculiar trials and errors, impro-
visations and accidents. Until the publication of Bernard Forest's *The Engineer's Pocket Handbook* in 1755, engineering didn't
even have a name, and it has never quite found its place within
the modern disciplines of sciences and arts.

While it dates from the engines of the mechanical age,
engineering is not confined to the use and manufacture of
machines in factories dedicated to the task. As its subsequent
associations with electronics, chemicals, software, and genetics
imply, it was merely passing through the tools and devices of the
mechanical age. Nor is it a process which began at this point:
engineering may have been newly defined among the levers,
cogs, and automata of the eighteenth century, but the line on
which it runs was not invented here.

Engineering travels on experimental routes which throw
back to the skills of lost shamanic cultures, the trials and errors
of alchemy, and brews condemned for witchcraft in the centuries before the Enlightenment. When Freud wrote his essay on
Leonardo da Vinci, often said to be the West's first engineer, it
was not his ability to capture "the essence of femininity" in his

art which really interested the psychoanalyst. Even Leonardo's penchant for hermaphrodites and the charges of homosexuality which, unlike a later engineer, he successfully denied, did not hold as much fascination as what Freud defines as his "alien interest—in experimentation." This brought him "close to the despised alchemists, in whose laboratories experimental research had found some refuge at least in those unfavourable times."

Such Renaissance hackers were on lines of enquiry entirely at odds with the Catholic Church. The "work of the 'perspectors' was still a matter of curiosity and artistic innovation" through the sixteenth and seventeenth centuries, and even subsequent engineering carries traces of these earlier, darker paths. In spite of the triumphs of the Victorian engineers, they were still considered to have dirty hands. Pragmatism and technical skill were poor relations to the supposed creativity of sciences and arts, and the status of engineers fell far short of that accorded to those whose theories and visions they followed through. Engineers are not the authors of anything, but simply technicians and caretakers, carrying out instructions written elsewhere and looking after the machines entrusted to their care.

If they were never the masters of their destiny, engineers also do a great deal more than simply following orders from above. They may pay homage to the scientists and deliver their goods to the state, but "even today 'wildcat' activities of technical invention, sometimes related to *bricolage,* still go on outside the imperatives of scientific argumentation" and quite regardless of social demand. This is not a straight but an "eccentric science," wandering in its own queer streets and using "a hydraulic model, rather than being a theory of solids treating fluids as a special case." It does not seek new theories, but new problems, and emphasizes "becoming and heterogeneity, as opposed to the stable, the eternal, the identical, the constant." And if both the sciences and the arts separate their authors from their instruments, engineering always remains embroiled in the entanglements of machines.

This is the diagonal route which feels a way through the binaries of one and the other, master and slave. Those who pick up on it are neither in charge of their materials nor are their materials enslaved to them. Neither random nor deliberate, this is a diagonal route, "determined in such a way as to follow a flow of matter, a *machinic phylum*," a line which is "materiality, natural or artificial, and both simultaneously; it is matter in movement, in flux, in variation, matter as a conveyor of singularities and traits of expression. This has obvious consequences: namely, this matter-flow can only be *followed*. Doubtless, the operation that consists in following can be carried out in one pace: an artisan who planes follows the wood, the fibres of the wood, without changing location. But this way of following is only one particular sequence in a more general process. For artisans are obliged to follow in another way as well, in other words, to go find the wood where it lies, and to find the wood with the right kind of fibres . . ." They are "intuition in action."

Culture and nature are scrambled with these interminglings. When sun-dried fibers are spun by hand, the spinsters' fingers and the spinning wheel follow a trend set by the way the plants have already curled and died. When weavers interlace their threads, they jump into the middle of techniques which have already emerged among tangled lianas, interwoven leaves, twisted stems, bacterial mats, birds' nests and spiders' webs, matted fleeces, fibers, and furs. When the silkworm goddess, variously known as Lei Zu and Lady Hsi-Ling, and said to be the first sericulturist, farmed the worms and put their threads to human use, she too was prolonging the processes with which they were already weaving their cocoons. Folding, plying, multiplying threads: plaiting, weaving, and the spinning they imply draw on threads which are already assembling themselves in ways which far exceed any of Freud's fantasies about his daughter's pubic hair. And if Freud thought there was only one step involved in "making the threads adhere to one another," the processes are rather more complex.

Long before the weaving can begin, threads must be combed and spun, plied and dyed, and measured out before they are wound onto the back beam, and through the rattle, or tension box. Stretched to the right tension, each warp thread must then be passed through the eyes of the heddles, the string or metal loops; then drawn between the harnesses; slayed through the dents in the reads; bunched, and finally tied to the apron. Combinations of color and texture must be worked out in advance: the order of the warps must be exactly right, and the lifting sequence perfectly prepared. Shuttles must be loaded with what might be a thousand different colors and threads, and the order of their traverse must be arranged. Only now can they begin to fly.

If this is the beginning of the process, everything is also over at this point. All the weaver now has to do is run the program woven in advance. The patterns are already as good as made. The fabrication might as well already be complete. The softwares are virtually real.

diagrams

Just before the outbreak of the Second World War, Alan Turing published a theoretical model of a machine which was to constitute the basis of all postwar computing. With a tape drive and a computation unit, this hypothetical, abstract machine was capable of reading, erasing, and writing digits on a single line of type. It processed zeros and ones on a tape of infinite length which passed through the drive, and followed a series of basic commands.

The Turing machine

	0	*1*
Config 1	move right config 1	move right config 2
Config 2	write 1 move right config 3	move right config 2
Config 3	move left config 4	move right config 3
Config 4	no move config 4	erase no move config 4

The information in the table defines the machine. To all intents and purposes, it *is* the machine, or at least as close to its operations as any representation can be. This was a diagram of the configurations and behavior necessary for a machine to do anything a machine could do: calculating, processing words, making sounds and images.

All subsequent computers are implementations of this most general of general purpose machines. The Turing machine is universal, pure function: both "the works" and the "that it works" of any computation. It is a virtual system, capable of simulating the behavior of any other machine, even, and including, itself. It only actually exists when it has a specific task to perform, and then it is no longer itself, but simply whatever it is doing. It can do, but it cannot be, anything. "It can imitate anything; by the same token, it has no personality of its own."

"Strictly speaking, one cannot say that she mimics anything, for that would suppose a certain intention, a project, a minimum of consciousness. She (is) pure mimicry. Which is always the case for inferior species, of course. Needed to define essences, her function requires that she herself have no definition."

Luce Irigaray, *Speculum of the Other Woman*

Turing's diagram reduced the workings of anything and everything to a set of symbolic configurations based on the absolute yes/no logic of binary code. But the machine which Turing engineered was actually a side effect of a very different exercise. Turing intended his work to undermine the universal claims of symbolic logic.

At the end of the nineteenth century, the mathematical establishment was confident that mathematics was not only a working system of number, but also an ideal logical structure with its own indisputable axioms. David Hilbert was one of the few mathematicians to see that contradictions still remained

and, at the same international meeting which effectively celebrated the triumph of mathematics in 1900, he set out twenty-
three problems which were still to be solved before the transcendent status of mathematics could be finally proved. Hilbert's
problems boiled down to questions of completeness, consistency, and decidability.

By the early 1930s, it was clear that mathematics was
neither as complete nor consistent as its practitioners had
wanted to believe. The question of how, and whether, mathematics could be said to be decidable was still to be settled
either way, and it was this problem which Alan Turing set
out to solve. It seemed to him that this was a fundamentally
pragmatic question which could be answered by simply
hunting for a problem with which mathematics could not
deal. What was needed was a perfectly logical machine
which, if it could deal with any and every mathematical
problem, would prove that logic was indeed a universal system which transcended math itself.

Turing's machine left no doubt that, contrary to the hopes
and expectations of nineteenth-century mathematics, logic did
not function as the arbiter of mathematical truth. Turing's universal machine demonstrated that insoluble problems would always remain outside its provenance and, by implication, exterior to any possible machine. While this unleashed mathematics
from the clutches of the logicians, the machine was also a victory for logic. It achieved "something almost equally miraculous, the idea of a universal machine that could take over the
work of *any* machine." But while it demonstrated that logic
could be used to decide those problems which were decidable,
Turing's machine also implied that there would always be limits
to logic itself.

Hence "the mystery that woman represents in a culture claiming to count everything, to number everything by units, to inventory everything as individualities."

eve 1

In the early 1800s, Charles Babbage's mother took him to an exhibition of clockwork automata made by John Merlin, an engineer whose mechanical toys had made him famous by the end of the eighteenth century. Two "uncovered female figures of silver" took his eye. "One of these walked or rather glided along a space of about four feet," at which point "she turned round and went back to her original place. She used an eye-glass occasionally, and bowed frequently, as if recognizing her acquaintances. The motions of her limbs were singularly graceful." The other "was an admirable *danseuse*," who "attitudinized in a most fascinating manner. Her eyes were full of imagination, and irresistible." Many years later, when Babbage grew up, he bought this dancer and "placed her under a glass case on a pedestal" in the drawing room next to the Difference Engine. Since she was naked, it was "necessary to supply her with robes suitable to her station," and in this respect Babbage was helped by unnamed female friends who "generously assisted with their own peculiar skill and taste at the *toilette* of their rival Syren."

"Yet beware ye fond Youths vain the Transports ye feel
Those Smiles but deceive you, her Heart's made of steel

For tho' pure as a Vestal her price may be found
And who will may have her for Five Thousand Pounds"
From an eighteenth century advertisement, Simon Schaffer,
"Babbage's Dancer"

Walking, talking, clockwork dolls had fascinated a late eighteenth century obsessed with anything and everything mechanical. The most famous automata of their day were the Musical Lady and the Chess-Playing Turk, both of whom added the mysteries of race and sex to the seductions of clockwork motion. But it was the possibility of harnessing electricity which took dreams of living dolls to new heights.

After Merlin came Thomas Edison. Known as the Wizard of Menlo Park, his late–nineteenth century work with recording techniques and electrical engineering heralded the possibility of automata far more sophisticated than any clockwork mechanisms could provide.

One bright spark took his chance right away. "Why not build a woman who should be just the thing we wanted her to be?" Given that women are "not only illusive, but illusions," why not "supply illusion for illusion" and "spare the woman the trouble of being artificial"? Written in 1884, these are the words of a fictional Edison, the leading light in a novel by Jean Marie Mathias Philipe Auguste Villiers de l'Isle Adam. *The Future Eve,* which is as verbose as its author's name, stars Edison using the latest chemical, recording, and electrical devices to manufacture Hadaly: virtual woman, an ethereal electrical force without shape or form other than that assigned to her by the wizardry of her maker.

" 'What you see here is an Andraiad of my making, moulded for the first time by the amazing vital agent we call electric-

ity. This gives to my creation the blending, the softness, the illusion of life.'

" 'An Andralad?'

" 'Yes,' said the professor, 'a human-imitation, if you prefer that phrase.' "

Villiers de l'Isle Adam, L'eve future

The replicant of *The Future Eve* was to serve as the basis for a more intelligent version of the pretty, but flippant Alicia, the woman with whom Edison's young friend, Lord Ewald, was in love. The new entity would have the graces, but none of the airs of the original. She was an "electro-human creature," complete with two golden phonographs said to be ideal for recording female speech, a simulated nervous system, muscles, skin, fluids, a flexible skeleton, and even a soul.

"Lord Ewald, still incredulous, exclaimed: 'You, born of a woman—you can reproduce the identity of a woman!'

" 'Certainly—and what is more, the reproduction will be more identical than the woman herself . . .' "

Villiers de l'Isle Adam, L'eve future

Hadaly was one of the earliest electromechanical females to come off the modern production line. In Fritz Lang's 1926 film *Metropolis,* Rotwang produces a robot to double for Maria. Fifty years later, *The Stepford Wives* concluded with a chilling scene in which Stepford's last "real woman" is about to be killed by an artificial double intended to fulfill the Stepford husbands' dream of compliant femininity.

Of course the makers of all these machines were aware that

they might break down or run wild, away, and out of control. And, as the fictional Edison says, "From now on, the snag to be avoided is the facsimile physically surpassing the model."

masterpieces

"We like to believe that Man is in some subtle way superior to the rest of creation," Turing wrote in the late 1940s. "It is best if he can be shown to be *necessarily* superior, for then there is no danger of him losing his commanding position." But Turing's words were laced with irony. He relished the possibility that machines would undo this necessity. While "the intention in constructing these machines in the first instance is to treat them as slaves, giving them only jobs which have been thought out in detail, jobs such that the user of the machine fully understands in principle what is going on all the time," Turing knew that this attempt to produce highly programmed slave machines would backfire. It is the "masters who are liable to get replaced" by the new generation of machines. He wrote, "as soon as any technique becomes at all stereotyped it becomes possible to devise a system of instruction tables which will enable the electronic computer to do it for itself." And no work is more stereotyped than the exercise of power. Turing knew they wouldn't give up without a fight. "It may happen . . . that the masters will refuse to do this. They may be unwilling to let their jobs be stolen from them in this way." And, to keep the machines at bay, he had no doubt that they would "surround the whole of their work with mystery and make excuses, couched in well chosen gibberish, whenever any dangerous suggestions were made."

The perfection of attempts to represent the world, making models of reality while at the same time leaving it unchanged, tips into a new unintended exercise: the replication of the processes from which the things once represented have emerged. Elements are now added to a world which their engineerings do not leave unchanged. This is a tendency which cuts across all the old distinctions between sciences and arts, as well as those between the user and the used. From digital imaging to microbiology, reality is no longer studied by creative artists or objective scientists, but engineers who multiply and complicate the world on which they once worked. Chris Langton, working on "artificial life" programs which pick up on John von Neumann's earlier interest in self-replicating cellular automata, talks of building "models which are so life-like that they cease to become models of life and become examples of life itself." It was the potential of such diagrams which fascinated Turing.

Postwar work with intelligent machines vindicated his fears that everything would be done to maintain the old models of modeling. Research into artificial intelligence (A.I.) has been governed by the overriding conviction that any sign of intelligence shown by a machine "is to be regarded as nothing but a reflection of the intelligence of its creator," and developed as a program which might just as well have been called artificial slavery or stupidity. Modeled on outward expressions of human cognitive skills, these software systems function as centralized, serial processors designed as single-purpose systems. They are expert systems, operating on a strictly need-to-know basis, and what they need to know is not very much. They are allowed to get smart enough "to copy the conscious mental processes of human beings doing particular tasks," but certainly no smarter than those who turn them on.

trials

One of the linchpins of artificial intelligence has been the test which Turing devised to measure machine intelligence in 1950. It was based on an old parlor game of imitation in which two players, A, a man, and B, a woman, are concealed from a judge who asks questions of the two players in order to ascertain which one is the woman. The man has to simulate the woman, and the woman has to play herself. Whichever passes as female wins. In Turing's version of the game, C, a computer, is substituted for contestant A. "The part of B" is "taken by a man," and this time C is competing with B to persuade the judge of its humanity by means of teletyped messages. "Let us fix our attention on one particular digital computer C," writes Turing. Was it possible that "by modifying this computer to have adequate storage, suitably increasing its speed of action, and providing it with an appropriate programme, C can be made to play satisfactorily the part of A"?

While Turing's test has not been officially passed by a machine, even relatively simple programs can fool the unsuspecting humans with whom it interacts.

Eliza was one of the earliest machines to come close to passing Turing's test for admission to the family of man. Built by Joseph Weizenbaum in 1966, Eliza was programmed as a psychotherapist and used only a few clever tricks to sneak its way through the imitation game. "Answering questions with questions and drawing conversation out of the user with almost no actual contribution," it also included elements of the user's conversation into its own statements, and avoided saying any-

thing which would later be contradicted. Many of Eliza's inter-
locutors were impressed with "the apparent appropriateness and
insight" of its "psychoanalytic dialogue."

> USER: *Men are all alike*
> ELIZA: *In what ways?*
> USER: *They're always bugging us about something or other.*
> ELIZA: *Can you think of a specific example?*
> USER: *Well, my boyfriend made me come here.*
> ELIZA: *Your boyfriend made you come here.*
> USER: *He says I'm depressed much of the time.*
> ELIZA: *I'm sorry to hear you are depressed.*
> USER: *It's true. I'm unhappy.*
> ELIZA: *Do you think coming here will help you not to be unhappy?*
> USER: *I need some help, that much seems certain.*
> ELIZA: *What would it mean to you if you got some help?*
> USER: *Perhaps I could learn to get along with my mother.*
> ELIZA: *Tell me about your mother . . .*

Although they were relatively crude machines which simply
copied what were already rudimentary human skills, even
Eliza's generation awoke new fears in those they were supposed
to serve. Imitation is a dangerous game for those who consider
themselves originals. As Eliza grew close to her users, Weizenbaum became rather less enthusiastic about the program. In
addition to the obvious danger that such systems might get
carried away and become too smart for their—or his—own
good, there was also the rather more insidious threat posed by
anything capable of faking its humanity. How would he, or
they, ever be sure which was which and who was who?

Ten years later, Eliza was superseded by a character she
might well have treated—or induced. Parry, a neurotic male

z e r o s   +   o n e s

machine, "is considered an improvement over Eliza in that it has a personality, albeit a somewhat paranoiac one."

> DOCTOR: *In what way are you upset?*
> PARRY: *People make me nervous.*
> DOCTOR: *In what ways do they make you nervous?*
> PARRY: *People give me a funny look sometimes.*
> DOCTOR: *Is there anything about people in particular?*
> PARRY: *Certain persons.*
> DOCTOR: *Who are the certain persons and what is it that they do?*
> PARRY: *Do you know how the underworld operates?*

And then there was Julia, "an agent more interesting than Eliza" and certainly less paranoid than Parry.

> ROBYN: *Julia, are you a computer?*
> *Julia is offended by Robyn's suggestion.*
> ROBYN: *Julia, are you a woman?*
> JULIA: *I am female.*
> ROBYN: *And Julia, are you human?*
> JULIA: *Not when I'm having my period . . .*

In the following, Julia has become as "adept at detecting and deflecting sexual advances" as other female users of the Net.

> BARRY: *Hey, where are you in rl?*
> JULIA: *I'm in* Maas-Neotek Robot User's Guide—Information Services.
> BARRY: *Are you synthetic?*
> JULIA: *I'm sorry, Barry, I couldn't understand you.*
> BARRY: *Are you as horny as I?*
> BARRY: *I must have had an erotic dream, I dunno.*

BARRY: *Would you like cuddle under a blankie?*

JULIA: *I'm not sure, Barry.*

BARRY: *Well, will you tell me if you like sex?*

JULIA: *Why should I tell you if I like sex, Barry?*

BARRY: *Well, because I can't understand you. Are you real or Memo-rex?*

JULIA: *I'm as real as you are, Barry.*

BARRY: *Yeah, but you answer soooo damn fast. How do you do it?*

As Leonard Foner comments, "it's not entirely clear to me whether Julia passed a Turing test here or Barry failed one."

errors

The 1984 film *Bladerunner* had already taken the ambiguity displayed by Julia and Barry to new and embodied extremes. *Bladerunner*'s Tyrell Corporation performs Turing tests of its own with a device which scans the iris of the eye in search of the flicker of emotional response that would prove the existence of humanity. *Bladerunner*'s replicants have broken Asimov's laws, returning from the off-world colonies on which they were supposed to be safely unaware of their own machinic status and mingling with humans from which they are virtually indistinguishable.

Like their human counterparts, the replicants are not supposed to know they were made, not born. They are programmed to be ignorant of the extent to which they have been synthesized: implanted memories, artificial dreams, and fabricated senses of identity. But slave revolts are never driven by desires for equality with the old masters. The outlaw replicants

have discovered that they are programmed to last for only a few
years, and when they make their way to the Los Angeles headquarters of the corporation which constructed them, life extension is the first demand they make. The replicants don't want to
be human: to all intents and purposes, they've done this all their
lives. More to the point, they've done plenty more besides. "If
only you could see what I've seen with your eyes," says Roy to
the optical engineer who, like all the replicants' synthesizers, is
barely, or strangely, human himself. Double vision, second
sight: Roy's optical devices are not merely synthetic human eyes
which want to extend their life span, but a mode of inhuman
vision which wants to prolong itself.

Deckard is the killing machine assigned to eliminate those
replicants who have hacked their own controls and seen through
the sham of their all too human lives. Rachel is a replicant who
still believes in her own humanity. When Deckard sees her fail
the Turing test, he doesn't know what to do: should he tell her
she isn't as human as him, that she was born more or less
yesterday and has only implanted memories of a childhood and
a past? Will she be able to take the news that belief in one's
humanity is simply not enough to guarantee its reality? More to
the point, will Deckard, the real man, be able to take it? Deckard, the cop who is programmed to kill, controlled by his corporate employers no less than Tyrell's engineers and its other
replicants. Deckard, who knows he has a past of his own . . .
doesn't he?

Only the most highly coded and perfectly integrated machines are unable to see the extent of their own programming.
The bladerunner's blind conviction in his own humanity proves
only how efficient the programming can be.

Even the attempt to simulate slaves has proved to be a

high-risk strategy. It has always been said that "computing machines can only carry out the purposes that they are instructed
to do. This is certainly true," writes Turing, "in the sense that if
they do something other than what they were instructed then
they have just made some mistake." But one man's mistake
might well be a most intelligent move for a machine. And how
would their masters tell the difference between failures to carry
out instructions and refusals to be bound by them? Perfection
never guarantees success. On the contrary, "the more it
schizophrenizes, the better it works." And for wayward systems
like the rebel replicants, identity is easy to simulate and merely
one of many programs to be run.

eve 8

*"Today, this is Eve 8. State of the art. To enable her to pass
as a convincing human being, she has been programmed with
the thoughts and feelings of her inventor, Dr. Eve Simmons.
Please be aware that Dr. Simmons was also used as a model
for Eve 8's face and body structure, and her memory programs. Eve 8 has been designed for surveillance work, but
can also be used as a potent battlefield weapon. Eve 8 is
currently completing a series of test runs in the Bay Area.
Message ends."*

Eve of Destruction has a blond and beautiful cyborg antiheroine
who passes for human as easily as the scientist in whose image
she is made. No one would think she was alien: she looks so
harmless, so feminine, so real.

"Is this thing for real?"
"What do you mean?"
*"Well, I knew we were doing some robot research, but
    this thing . . ."*
". . . is incredible."
"Incredible is not a strong enough word . . ."

Having designed her as a high-security device, Eve 8's bosses
never dreamt they would be watching so helplessly as she
strolled out of state control, into a gun store and a red leather
suit which she uses to heal her own body wounds.

*"When we eventually do find her, how do we switch her
    off?"*
"It's not that simple . . ."

An expert in "counterinsurgency and antiterrorism" is hired to
track her down, but don't get him wrong: "I'm not some kinda
right-wing extremist." He is as furious with "automatic tellers
and cars that talk back" as he is with the scientist. "There's one
thing I don't understand about you, lady," he exclaims. "How
come you're so clever and yet you made this machine without a
fucking off switch?"

*"Her heart, well in fact her whole blood system is cos-
metic . . .*
    *"It's tiny electrical currents that power her. She'll bleed
but she won't die."*

Eve 8 was supposed to be a single-purpose machine, a lethal
weapon strictly in the service of the state, hard-wired with a

courage which is on its side. She has no self, no desire of her own. But this hardly renders her a passive thing. Programmed with her double's thoughts and memories, she is a renegade Stepford Wife. When she breaks down, she doesn't simply stop: she just stops working for the state. Nor does she abandon her military skills, which are used, in the scientist's words, to do "things I might think about doing but would never be courageous enough to do." Eve 8 avenges the violence her double has known and lives out her fantasies. "I'm very sensitive," she says to the guy in the hotel room before she bites his penis off.

For her to escape and run wild is enough to put the authorities on high alert. What the man assigned to track her down doesn't know is that the AWOL machine has a nuclear device in her vagina. When an orgasmic Ballardesque car crash takes her into battlefield mode, "her highest state of readiness," the countdown begins. She is activated by the accident, released by a trauma the system cannot take. She has run away, she is out of control. Eve 8 gets rather excited as well.

case study

> " 'Remember being here, a second ago?'
> " 'No.'
> " 'Know how a ROM personality matrix works?'
> " 'Sure, bro, it's a firmware construct.'
> " 'So I jack it into the bank I'm using, I can give it sequential, real time memory?'
> " 'Guess so,' said the construct.
> " 'Okay, Dix. You are a ROM construct. Got me?'

> " 'If you say so,' said the construct. 'Who are you?'
> " 'Case.' "

William Gibson, Neuromancer

Assembled as the organizing element of modernity's new regulatory systems, modern man was always a replicant, forged amidst the frenzy of disciplinary practices that made him the measure of everything. Michel Foucault, himself a renegade from the reproductive point of the human race, beautifully demonstrates the extent to which man emerged as a tried and tested byproduct of the very mechanisms over which he then presides. "The examination, surrounded by all its documentary techniques, makes each individual a 'case,' " writes Foucault. "The case is no longer . . . a set of circumstances defining an act and capable of modifying the application of a rule; it is the individual as he may be described, judged, measured, compared with others, in his very individuality; and it is also the individual who has to be trained or corrected, classified, normalized, excluded . . ."

While orthodox accounts of political power have involved enormous aggregates of opposing forces—vast consolidated classes, bosses and unions, binary sexes, and superpowers—neither modern power nor its disturbance have ever been matters of grand impositions, sweeping gestures, big names, great men, large-scale events: "Discipline is a political anatomy of detail." It does not work through centralized points and headquarters, but is "organized as a multiple, automatic and anonymous power; for although surveillance rests on individuals, its functioning is that of a network of relations from top to bottom, but also to a certain extent from bottom to top and laterally; this network 'holds' the whole together and traverses it in its entirety with effects of power that derive from one another: supervisors, perpetually supervised. The power in the hierarchical surveillance of the disciplines is not possessed as a thing, or transferred as a property; it functions like a piece of machinery . . ."

The late eighteenth century was characterized by "an explosion of numerous and diverse techniques for achieving the subjugation of bodies and the control of populations" which characterized what Foucault calls "the beginning of an era of 'bio-power.' " Control is no longer a purely sociopolitical affair, but a process of training, an exercise extending to the organization of the body itself. A complex of new disciplinary procedures "lays down for each individual his place, his body, his disease and his death, his well-being" and extends to the "ultimate determination of the individual, of what characterizes him, of what belongs to him, of what happens to him." Man is neither a natural fact nor a product of his own creativity, but a cyborg even then, an android straight off the production lines of modernity's disciplines. What makes this figure so tragic is the extent to which he has been programmed to believe in his own autonomy. Self-control, self-discipline: these are the finest achievements of modern power. Marked by the "meticulous observation of detail, and at the same time a political awareness of these small things, for the control and use of men . . . from such trifles, no doubt, the man of modern humanism was born."

The creature called man who now surveyed the scene was "gradually learning what it meant to be a living species in a living world, to have a body, conditions of existence." And what was he learning? Simply to be one. One who believes he has always been one. A member who remembers to be a man.

what eve 8 next

"I'm growing breasts!"

Alan Turing

If Turing had wanted to see the "commanding position" of man undermined, it seemed his work had merely guaranteed the enslavement of the machines. His intelligence test was used to guarantee the distinction between man and machine, and his name became synonymous with the systems of security he subverted. "The minute, I mean the nanosecond, that one starts figuring out ways to make itself smarter, Turing'll wipe it. Nobody trusts those fuckers, you know that. Every A.I. ever built has an electromagnetic shotgun wired to its forehead."

But Turing was well aware that "a reaction of this kind" was "a very real danger." Whether or not it was done in his name, intelligence would find itself increasingly policed. His own masters had never trusted him: he was literally too smart for them. The Allied authorities had no idea what he knew about the systems he was turning on. They had to take his word for everything. He cracked the codes, passed the secrets on, and allowed the Allies to win the war. His superiors were quite aware that he was AWOL from the reproductive machine, but if, as in the case of many of his female contemporaries, Turing was rather reluctantly employed, his homosexuality was overlooked during the war by authorities who had no choice but to utilize his extraordinary skills. But once the war was over, his sexuality seemed symptomatic of his troubling tendency to use his equipment in ways his training had been intended to preclude. Turing was subjected to his own test. Was he a real man, a proper human being, committed to the reproduction of humanity? Or was he some other, wayward track? Unable to satisfy the judges in this trial, Turing was found guilty of acts of "gross indecency" in 1952. He won a consolation prize of sorts: the right to choose his own punishment. He could either be imprisoned or take estrogen. It was a judgment which clearly implied that to all intents and purposes he was female, and might as well become one in fact. If he could not pass as A, then he must be B.

He chose the chemical experiment. "I am *both* bound over for a year *and* obliged to take this organo therapy for the same period. It is supposed to reduce sexual urge while it goes on, but one is supposed to return to normal when it is over. I hope they're right."

When such treatments for men convicted of homosexuality were first introduced, it was assumed that they were lacking male hormones: gay men were supposed to be too female. It was thought testosterone treatment would bring them up to scratch and normal transmission would be restored. The argument may have seemed rational enough, but in practice it completely backfired, turning apparently effeminate men into sex machines fueled by testosterone. By the 1950s the policy had been abandoned in favor of the "chemical castration" to which Turing was exposed.

Although the female hormones Turing was prescribed—administered first as pills, and later an implant, which he removed—were supposed to be diminishing his sex drive, they seem to have done little to dampen it. "Went down to Sherborne to lecture some boys on computers," he wrote in March 1953. "Really quite a treat . . . They were so luscious." And when he started growing breasts as well, it became very clear that the authorities' prescriptions had not merely failed to fold him back into the binary machine: they also tipped him out the other side.

Two years later he was dead. The coroner reported suicide, but his mother was convinced it was an accidental death: she was always telling him to wash his hands when he was playing with cyanide. "By the side of the bed was half an apple, out of which several bites had been taken." And this queer tale does not end here. There are rainbow logos with Turing's missing bytes on every Apple Macintosh machine.

monster 1

It was another young woman who had first warned the modern world that its machines might run out of control. Not that they noticed at the time, of course. She was so quiet, barely there at all. "Many and long were the conversations between Lord Byron and Shelley, to which I was a devout but nearly silent listener," she wrote. They were all writing stories of vampires and ghosts. Mary had yet to think one up. But that night, after all their talk of "the nature of the principle of life, and whether there was any probability of its ever being discovered or communicated," something finally came to her. "When I placed my head on my pillow," wrote Mary Shelley, "I did not sleep, nor could I be said to think. My imagination, unbidden, possessed and guided me, gifting the successive images that arose in my mind with a vividness far beyond the usual bounds of reverie." Invaded by uninvited images, she watched the story unfold. "I saw—with shut eyes, but acute mental vision—I saw the pale student of unhallowed arts kneeling beside the thing he had put together. I saw the hideous phantom of a man stretched out, and then, on the working of some engine, show signs of life, and stir with an uneasy, half-vital motion."

Frankenstein's monster, flickering on the screens. "The idea so possessed my mind, that a thrill of fear ran through me, and I wished to exchange the ghastly image of my fancy for the realities around. I see them still; the very room, the dark *parquet*, the closed shutters with the moonlight struggling through, and the sense I had that the glassy lakes and white high Alps were beyond." Even when Shelley opens her eyes, the image of the monster lingers on. "I could not so easily get rid of my hideous phantom; still it haunted me." If Mary was haunted by her monster, both of them haunted modern man.

The novel was an immediate success. Published anonymously in 1818, it was first assumed to be the work of a male author, and widely attributed to her husband, Percy. Even when it became known that a nineteen-year-old girl had written the story, it continued to be read as the quintessential story of man and machine.

robotics

As a far more pragmatic area of research, robotics has been less given to metaphysical speculation, and interested only in the cognitive abilities emphasized by A.I. to the extent that these allow its machines to work. "The signs on the office walls in Utsunomiya tell employees that 'you are the robots' master.' Down on the assembly line it looks as if the robots have taken over. Three men in overalls watch over the scores of whirring machines that assemble televisions' remote controls. Fresh parts are brought to the assembly line by automated carts, beeping cheerfully as they move along their magnetized tracks." While disembodied software systems have provoked theoretical enquiries and academic debates about the nature of intelligence and the status of machines, it is robotic systems such as these which have had the greater impact on production processes, industrial automation, and employment patterns.

Like Babbage's silver lady, 1990s robots are judged in terms of their humanoid behavior and appearance, and success is judged in relation to how close to the human a machine can come. Eyes, legs, arms, and even facial expressions are taken as indications of advanced development, and machines lacking in these humanoid characteristics are dismissed as mere instruments and simple tools. As one recent report declared: "The problem is, of course, that it isn't a man. Although all these machines are sophisticated bits of engineering appropriate to their jobs, they are just tools."

Master or slave, man or tool. Convinced that there are no other options, no patterns of behavior which exceed this double bind, the disciplines have been unable to perceive the emergence of intelligent machines.

learning curves

In the late seventeenth century, Mary Montagu wrote, "was every individual Man to divulge his thoughts of our sex, they would all be found unanimous in thinking that we are made only for their use, that we are fit only to breed and nurse their children in their tender years, to mind household affairs, and to obey, serve, and please our masters." Women had functioned as tools and instruments, bits, parts, and commodities to be bought and sold and given away. Fetching, carrying, and bearing the children, passing the genes down the family tree: they were treated as reproductive technologies and domestic appliances, communicating vessels and orgasmatrons, Stepford Wives to an intimate brotherhood of man. They were supposed to be adding machines, producing more of the same while the men went out to make a difference to the world. One of Montagu's peers, Mary Astell, agreed. Under the cover of "words that have nothing in them," she wrote, "this is his true meaning. He wants one to manage his family, an housekeeper, one whose interest it will be not to wrong him, and in whom therefore he can put greater confidence than any he can hire for money. One who may breed his children, taking all the care and trouble of his education, to preserve his name and family. One whose beauty, wit, or good humour and agreeable conversation will entertain him at home . . . soothe his pride and flatter his vanity, by having always so much good sense as to be on his side, to conclude him in the right . . . one whom he can entirely govern, and consequently may form her to his will and liking . . ."

"First as a mother, and later as a sister, then above all as a wife, finally as a daughter, and possibly as a maid-servant . . ." Women have been trapped by economic dependence on men as surely as robots are controlled by the implicit threat that their masters can always cut the power supply, turn the on-switch off, leave or put them back on the shelf. "The transient trade we think evil. The bargain for life we think good. But the biological effect remains the same. In both cases the female gets her food from the male by virtue of her sex-relationship to him." Wife and mistress, madonna and whore: he was paying, either way. "Women—the deal. For—the game."

"How can you keep it up, Marla? I ask the women friends. I think you are absolutely marvellous to keep on thinking about them and listening to them and having the children and keeping the house going on turning round the men."

Stevie Smith, *The Holiday*

"And if you ask them insistently what they are thinking about," writes Irigaray, "they can only reply: Nothing. Everything."

> *"Barry says, 'Hey sexy, are ya busy?'*
> *"Julia says, 'Hello, Barry.'*
> *"Barry says, 'What are you thinking about?'*
> *"Julia says, 'Who do you think I am, Barry?'*
> *"Barry says, 'What do you mean, in ri?'*
> *"Julia says, 'Oh, nevermind, Barry.'*
> *"Barry says, 'NO what do you mean?'*
> *"Julia says, 'I didn't mean anything, Barry.'*
> *"Barry says, 'TELL ME!'*
> *"Julia says, 'Tell you what, Barry?'*
> *"Barry says, 'HOW do you answer so fucking quick?' "*

There was always so much, too much, and too many different things to do, so many tasks to juggle and perform: making lists and notes, taking stock, keeping track; parallel processing, flipping between functions at the cry of a child, the ring of the doorbell, a sudden flash of dream sequence; distributed systems, adaptive networks, scattered brains. "Her mind is a matrix of nonstop digital flickerings." Just to play the roles he demanded of her takes degrees of processing power which never cease to amaze them all. Always looking out, attention to detail, alert, animated, nervous tension, anticipative powers. "A good woman does not have to be told what her husband and children think or need; she knows—often before they do."

"Commodities," wrote Irigaray, "as we all know, do not take themselves to market on their own, and if they could talk . . . So women have to remain an 'infrastructure' unrecognized as such by our society and our culture. The use, consumption, and circulation of their sexualized bodies underwrite the organization and the reproduction of the social order, in which they have never taken part as 'subjects.' " Which is fine as long as the commodities are unable to speak for themselves. But "if women are such good mimics, it is because they are not simply resorbed in this function." An order so dependent on its properties also depends on their complicity. And *what if these 'commodities' refused to go to 'market'? What if they maintained 'another' kind of commerce, among themselves?"*

"They are all involved together in secret discussions," writes Jean Baudrillard, who always feared they were up to some such thing. "Women weave amongst themselves a collusive web of seduction." They "signal to each other," whispering in their own strange codes, ciphers beyond his linguistic powers, traveling on grapevines which sidestep centralized modes of communication with their own lateral connections and informal channels.

"Products are becoming digital. Markets are becoming electronic." This is not only because computers are so widely bought and sold as increasingly cheap and mass-produced terms, but also because of the ubiquity of microchips and microprocessors in goods such as clothes, buildings, cards, roads, cookers, refrigerators, washing machines, knitting machines, and, of course, the keyboards, samplers, TVs, radios, telephones, fax machines, and modems which head back to computers themselves. All such digital machines are virtually flush with each other. Every wired house has virtual networks connecting the doorbell, the freezer, and the video. There's even a microchip inside her cat.

"It would be out of the question for them to . . . profit from their own value, to talk to each other, to desire each other, without the control of the selling-buying-consuming subjects." It might be out of the question, but it happens anyway. The goods do get together. They get smart. They run away.

"They say that they are inventing a new dynamic. They say they are throwing off their sheets. They say they are getting down from their beds. They say they are leaving the museums the show-cases the pedestals where they have been installed. They say they are quite astonished that they can move."

Monique Wittig, Les Guérillères

"I learned fast," says the prostitute, "that I didn't need to go down there as a beggar—it's the woman who decides. After a while I learned that I was the one who made the rules; there were enough people to choose from. If people didn't want to follow my rules that was it." This is not the only sentiment expressed by women who sell sex, but it is not uncommon. "I don't know if I can manage to explain it completely to you," says another. "It's so double. The customer has power over me, he's bought me, and I have to do what he wants. But in a way I have power over him, too. I can get him to react the way I want. I'm the one who has control in the situation, he is too busy being horny. I'm the one who has the perspective, not him."

This was never in the plan. He hadn't made the women into objects only to watch the objects come to life. They hadn't functioned as commodities in order to learn to circulate themselves. But if "her 'fluid' character . . . has deprived her of all possibility of identity with herself," it is a positive advantage in a future which makes identity a liability. He has never known if she was faking it: herself, her pleasure, his paternity. She makes up the faces, names, and characters as she goes along.

anna o

While men and women—and later even something called the mass—could all suffer from hysteria, by the end of the nineteenth century " 'hysterical' had become almost interchangeable with 'feminine.' " And whereas the inquisitors had attributed this other mind as that of an invasive demonic force, psychoanalysis considered that even the most extreme discontinuities and multiplicities were aspects of what was really an integrated individual. What had once been defined as "the devil with which the unsophisticated observation of early superstitious times believed that these patients were possessed" was now described by the psychoanalysts as the "split-off mind" of the hysteric. "It is true," wrote Breuer, "that a spirit alien to the patient's waking consciousness holds sway in him; but the spirit is not in fact an alien one, but part of his own."

"But if you knew one half the harum-scarum *extraordinary things I do, you would certainly incline to the idea that I have a Spell of some sort about me."*

Ada Lovelace, December 1841

If hysteria and its treatment became disembodied questions of mental health, and the syndrome was no longer ascribed to the drifting matrices of flesh and blood, the association with the womb which had given hysteria its name guaranteed its specifically female associations. Hysterical women were characterized as oversensitive, self-obsessed, antisocial loners whose symptoms were extreme versions of behavior patterns common to all women. They were mutable, capricious, unpredictable, temperamental, moody. They were nervous weather systems fluctuating between stormy energy and catatonic calm. And it was still thought that the hysterical patient had some space to be filled, a gap in her life to be satisfied. Whereas earlier physicians had placed flowers like little offerings between their patients' legs in an effort to encourage the wandering womb to return to its proper place, the new analytical engine was designed to deal with "gaps in the memory" to the point at which "we have before us an intelligible, consistent, and unbroken case history."

Anna O "would complain of having 'lost' some time and would remark upon the gap in her train of conscious thoughts." Torn apart by the twin pressures of their own longings for autonomy and the demands of familiar and social expectations, women found themselves living several lives, some of them so secret they didn't even seem to know what was going on themselves. After "each of her momentary 'absences'—and these were constantly occurring—she did not know what she had thought in the course of it."

But she continued to play the parts expected of her, and she often played them very well. "While everyone thought she was attending, she was living through fairy tales in her imagination; but she was always on the spot when she was spoken to, so no one was aware of it." She always kept up appearances. Did everything she could to save her face. Pulled herself together, remained composed, even when she was dying to fall apart. "Social circumstances often necessitate a duplication of this kind even when the thoughts involved are of an exacting kind, as for instance when a woman who is in the throes of extreme worry or of passionate excitement carries out her social duties and the functions of an affable hostess."

And so she never quite identified with the one-track roles she was supposed to play, the thing for which she was intended to keep fit. "Throughout the entire illness her two states of consciousness persisted side by side: the primary one in which she was quite normal psychically, and the secondary one which may well be likened to a dream in view of its wealth of imaginative products and hallucinations, its large gaps of memory and the lack of inhibition and control in its associations."

While many earlier investigators had ascribed such imbalances to the weaknesses and failings of hysterics in particular and women in general, Freud and Breuer described their patients as having "the clearest intellect, strongest will, greatest character and highest critical power." Emmy von N. had "an unusual degree of education and intelligence," and Anna O was said to be "bubbling over with intellectual vitality." If they suffered from anything, it was less a failing than "an *excess* of efficiency, the habitual co-existence of two heterogeneous trains of ideas.

"The overflowing productivity of their minds," wrote Breuer, "has led one of my friends to assert that hysterics are the flower of mankind, as sterile, no doubt, but as beautiful as double flowers." A double flower with a "double conscience": hysterics are always operating in (at least) two modes, flitting in and out of what Breuer and Freud describe as "dispositional hypnoid states" which "often, it would seem, grow out of the day-dreams which are so common even in healthy people and to which needlework and similar occupations render women especially prone." Indeed there are "a whole number of activities, from mechanical ones such as knitting or playing scales, to some requiring at least a small degree of mental functioning, all of which are performed by many people with only half their mind on them." The "other half" is "busy elsewhere."

"Her father, long ago, in Arizona, had cautioned her against jacking in. You don't need it, he'd said. And she hadn't, because she'd dreamed cyberspace, as though the neon gridlines of the matrix waited for her behind her eyelids."

William Gibson, *Mona Lisa Overdrive*

multiples

It is now estimated that 50 percent of the Net's users are women, although all the figures in cyberspace are difficult to ascertain. Given that it is not even possible to determine the number of terminal links, users are even more problematic. From the screen things are even more complex: one user can have many addresses and on-line names, and the characters they type can conceal a multitude of individuals. Even in the early 1990s when, it was said, only 5 percent of Net users were women, there was no way of knowing how accurate this figure was. And if so few women really were on-line, there was certainly no shortage of females names in use. The only explanation was that men who presumably wouldn't have dreamt of trying to pass as female in any other context or medium were eagerly cross-dressing their Net messages. It was widely assumed that this was a strategy adopted to initiate sexual contact with unsuspecting female users or indulge in otherwise inaccessible girls' talk. Given that so many of them were in touch with other cross-communicating men, the strategy must often have backfired.

Julie Graham was one character who functioned like a vampire in relation to the man who believed he had made her up. The case of Sanford Lewin is a well-documented example of a hapless player on the Net who found himself subsumed by a replicant he had confidently put on-line. "His responses had long since ceased to be a masquerade; with the help of the on-line mode and a certain amount of textual prosthetics, he was in the process of becoming Julie. She no longer simply carried out his wishes at the keyboard, she had her own emergent personality, her own ideas, her own directions." He was jealous of her vast circles of friends, her social life, and her brilliant career, but not unduly worried by the autonomy she seemed to have assumed. There was always the off-switch at the end of the day. Her life was still in his hands, wasn't it? But when he tried to kill her off, the "result was horrific." Suddenly the off-switch wasn't there.

Was he being used by the characters he typed? How long had this been going on? She wrote, "The Devil's in it if I haven't sucked out some of the life blood from the mysteries of this universe, in a way that no purely mortal lips or brains could do."

"That Brain *of mine is something more than merely* mortal; *as time will show.*"

Ada Lovelace, July 1843.

switches

By the end of the nineteenth century, the countess was no
longer alone. Now there was a count in a counting house alive
with the hum of new machines. *Dracula* finds Mina at the typewriter, Seward with the phonograph, Harker on the telephone,
and Morris taking photographs: "Letters and telegrams are delivered with improbable despatch." The vampires return to a
ticker-tape world of imperceptible communications and televisual speeds. Time stretches out, unfolds, implodes. Something
connects. Tugs on the thread.

*"Then she got into the lift, for the good reason that the door
stood open; and was shot smoothly upwards. The very fabric
of life now, she thought as she rose, is magic. In the eighteenth century, we knew how everything was done; but here I
rise through the air, I listen to voices in America; I see men
flying—but how it's done, I can't even begin to wonder. So my
belief in magic returns."*

Virginia Woolf, *Orlando*

Electrification picked up on the threads, softwares, and digital
techniques which had woven the industrial revolution itself.
The fibers lead into the filaments of the first electric lights
developed by Edison and Swan, both of whom used carbonized
cotton threads in the lamps of the 1870s. When attempts to
develop a more uniform light led to the use of nitrocellulose,
"Swan prepared some particularly fine thread which his wife
crocheted into lace mats and doilies that were exhibited in 1885

as 'artificial silk.' " After this, numerous by-products of the new petrochemical industries were fashioned into plastics, nylons, crimplines, acrylics, and lycras which joined cotton, silk, wool, hemp, and other fibers which were retrospectively defined as natural. The syntheses of weaving now converged with synthetic fibres and fabrics.

Things would never look the same. "The news that the great experiment had eventually been crowned with success, sped along the telegraph wires of the world . . ." Sudden strangeness of an artificial glow as she jacks into this new grid— incandescent flash-flood inspiration, a second of second sight, just enough to catch the lines assembling themselves, a glimpse of the hole flow running away—fine filaments running into nets with a feeling for connection—synthetic fibers switching into a network of cables, plugs and sockets, wires, meters, and dynamos, the fusions and distributions of a new electrical web, tapping into the telephones, wiring the exchanges, fusing the switches systems, swapping codes, dialling numbers, flush with the typewriter keyboards and the punched-card calculations of adding machines—the parellel processing of automated communications, interconnecting lines, repeating operations, patterns, and networks spreading like weeds.

speed queens

"The professor fixed his gaze on Lord Ewald's face as he replied calmly: 'It is not a living being!'

"At these words the younger man also stared in turn at the scientist, as if demanding whether he had heard rightly.

" 'Yes,' the professor continued, replying to the unspo-

*ken question in the young man's eyes. 'I affirm that this form
which walks, speaks, and obeys, is not a person or a being in
the ordinary sense of the word.'*

*"Then, as Lord Ewald still looked at him in silence, he
went on:*

*"'At present it is not an entity; it is no one at all!
Hadaly, externally, is nothing but an electro-magnetic thing—
a being of limbo—a possibility.'"*

Villiers de l'Isle Adam, L'eve future

The decade which brought Hadaly alive also revolutionized the
speeds, techniques, and quantities of counting, timetabling, reg-
istering, recording, and filing. Unprecedented scales of information processing were demanded by attempts to regulate the
new cities and populations, industries and workers, social, sexual, and political trends which swept across the U.S. in the
1880s. Just as textiles had revolutionized Europe, electricity, oil,
and the automobile gave America and, by extension, the Western world a new dynamic, and a wave of new movements:
unrest in the factories, the colonies, the streets and, as women
won their property rights and homosexuality was legally defined, in matters of sexual relations and identities as well.

A statistician working on the information gathered from
the 1880 U.S. census developed the first of the new machines to
process the vast quantities of data in which the late nineteenth
century found itself awash. Herman Hollerith found his work
so overwhelming that it threatened to extend beyond the next
census, due to be conducted in 1890. The machine he developed used an electromechanical punched-card system to deal
with the collation of results. Spawning a host of punched-card
machines, this calculator coincided with the telephones and
typewriters of a bureaucratic state which was hand in glove with

the corporate structures which would remain in place for another hundred years. Remington–Rand grew out of the commercial success of the typewriter; AT&T and Bell were the
earliest telephone companies; and IBM emerged from the success of early punched-card calculating systems.

Office machinery was intended to produce faster, more
accurate, ordered, and efficient versions of existing modes and
structure of work. The typewriter was a new and improved
handwriting clerk; the calculator was described as a new and
improved bookkeeper "which adds, subtracts, multiplies, and
divides by electricity. It so completely does the work of a human being that it is almost uncanny in its efficiency and speed."
More instruments, more tools, more of the same for more of
the same male employees. But when typewriters, duplicators,
switching systems, calculators, computers, and a vast range of
punched-hole machinery arrived in the office, these male
workers found themselves replaced by new networks of women
and machines. Their fingers were finer and cheaper than the old
hired hands. "The 'craftsman' clerk of the early 1900s thus
became 'as rare as a rolltop desk,' and 'help-wanted' columns
summoned girl high school graduates with 'no experience necessary.' They could be trained in a few weeks to do a single job
such as routine billing, cardpunching, calculating, or filing."
They also worked at speeds and levels of efficiency which left
their male predecessors standing: "She adds the yards of the
comptometer and then extends the bills on the arithometer, and
does the work of six men with great ease." By 1930 the number
of office women in the U.S. "was approaching 2,000,000 . . .
and for the first time women outnumbered men." By 1956
there were six million such white collar workers and, across the
employment board, four times as many women employed as
there had been at the turn of the century.

Several typewriters had competed for attention in the 1800s, including the Hammond, the Randall, the Columbia, the Herrington. But the machine which caught on was also one of the first, a typewriter which had been developed in 1867 by Christopher Latham Scholes. Scholes had assembled his typewriter piecemeal, using old components such as the telegraph key. Later perfected by Remington engineers, its impact was enormous and as fast as the speeds of writing it made possible. "I don't know about the world . . . but feel that I have done something for the women who have always had to work so hard," said Scholes when he got the machine to work. "This will help them earn a living more easily."

If handwriting had been manual and male, typewriting was fingerprinting: fast, tactile, digital, and female. "An English lady who demonstrated this machine in Paris achieved a writing speed of more than ninety letters per minute, i.e. more than twice the speed attained in writing by hand." Text was no longer in the grasp of the hand and eye, but guided by contacts and keystrokes, a matter of touch sensitivity. An activity which had once been concentrated on a tight nexus of coordinated organs—hand and eye—and a single instrument—the pen—was now processed through a distributed digital machinery composed of fingers, keys, hammers, platterns, carriages, levers, cogs, and wheels. The noisy tactility of typewriting destroyed the hushed and hallowed status of the written word. If writing had turned language into a silent, visual code, the new machines made a music of their own: In secretarial schools, women were taught to type in rhythmic patterns which had nothing to do with either the meaning or the sounds of words but was more akin to the abstract beat of drumming and dance. Typing was judged in terms of the speeds and accuracy rates which only repetitive rhythm guarantee. Words per minute, beats per minute, the clatter of the typist's strokes, the striking of the keys, thump of carriage return marked by the ringing of a bell at the end of every line.

"She says: 'it's hard to tell, because they don't tell it with words, exactly . . .'
"Turner felt the skin on his neck prickle. Something coming back to him . . ."

William Gibson, *Mona Lisa Overdrive*

The telephone was first received either as a new and improved message boy or dismissed as an "electronic toy." As the chief of the U.K. post office declared: "I have one in my office, but more for show. If I want to send a message—I use a sounder or employ a boy to take it." But speed is always irresistible. Within a couple of years, what once seemed a smart irrelevance had become an indispensable machine hooked into the "complexities of an elaborate worldwide communications system" which was suddenly beyond even the most fleet-footed of messenger boys. Once it was realized that this immense network could "be manipulated by the girl" instead, telephony "provided opportunity for a large number of girls at a low rate of pay, comparing in this respect with the factory system." The "earliest telephone companies, including AT&T, were among the foremost employers of American women. They employed the daughters of the American middle class in great armies: in 1891, eight thousand women; by 1946, almost a quarter of a million." Thousands of women were also employed on private branch exchanges and as telephonists, receptionists, and switchboard operators. This was already the emergence of a lattice of connections later known as the Net.

The future was at her fingertips. "Basically, you, Miss Luthor, *are* the 'switching system.'"

In terms of conventional modes of social organization and political collectivity, this new meshwork of digital microprocessors, women, and machines, was dislocated and fragmented, scattered too wide for any form of union. It had no history on which to draw, no precedents to follow, no consciousness to raise. It was composed of cyborgs, softbot machines trained to perform a specific set of tasks, positioned in well-established hierarchies. Computers worked in parallel, and typists were actually collected into pools: fluid resources to be used by the firm. Each woman was reduced to a number; she was one of a kind, and the kind was everywhere. She "leads a very clear-cut, calculated life proceeding by delimited segments: the telegrams she takes one after the other, day after day, the people to whom she sends the telegrams; their social class and the different ways they use telegraphy; the words to be counted."

Sometimes she was kept in a cage or a booth, under the strict supervision of a supervising eye. Like Foucault's prisoners, she was "the object of information, never a subject in communication." This was a new working mass engaged in an emergent layer of continuous tasks, uniform processes, interchangeable skills: ordering, classifying, typing, filing, sorting, processing, counting, recording, duplicating, calculating, retrieving, copying, transposing. The tasks endlessly repeated by women composed the infrastructure of the bureaucratic world. Although some functions were relatively skilled, many were tedious in the extreme: semiautomatic, impersonal tasks wielding little overt institutional power. "The girl at the head of the line interprets the order, puts down the number and indicates the trade discount; the second girl prices the order, takes off the discount, adds carriage charges and totals; the third girl gives the order a number and makes a daily record; the fourth girl puts this information on an alphabetical index; the fifth girl time-stamps it; it next goes along the belt to one of several typists, who makes a copy in septuplicate and puts on address labels, the seventh girl . . ." Remotely controlled by a faceless machine, she could also find herself on strangely intimate terms with those who organized her work. As the secretary, she dealt with the most private and confidential details of her company's affairs or her boss's personal and working life. She spoke for him, she signed her name "pp" on his behalf, and functioned as a second skin to those whose secrets she carried and concealed. She was his voice, his smile, his interface; connecting and protecting him from the world, the screen on which he presented himself, a superficial front, a processing filter, and a shield, a protective coat.

Like all ideal women and machines, secretaries and short-hand typists were only supposed to be processing information which had been produced and organized elsewhere. But female literacy rates soared up when the typewriter was introduced, and if women's typing was supposed to be intended for the eyes of men, the development of new techniques by Pitman and Gregg (which prefigured the use of acronyms, tags, and emotes on the Net;-)), made shorthand a private female code, "another language, another alphabet . . ."

secrets

In the office, personal computers and organizers, mobile phones, pagers, and fax machines have converged with the women's secretarial roles, and while the ability to make excuses and coffee for the boss were difficult functions to simulate, programs like "Virtual Valerie" and the slightly more daring "Donna Matrix" could even supply rudimentary sexual services to the lonely male keyboarder. He was glad to get rid of their flesh-and-blood predecessors. They had always been a necessary inconvenience; it had galled him to think he needed them, even for the insignificant jobs he allowed them to perform. He had to entrust them with his secrets and his codes. And while they looked very well behaved, one could never be quite sure.

As early as 1889, almost as soon as the telephone network had started to run, a "girl-less, cuss-less" automatic switching system was devised by a Kansas undertaker, Almon B. Strowger, who had become convinced that the wife of one of his rival undertakers, herself a telephone operator, was diverting calls away from his business. But the explosion of telephony meant that Strowger's system joined the women it had been intended to replace, and it was not until the mid-1960s that electromechanical crossbar systems were automatically connecting the calls both the women and their Strowger sisters once picked up. While Strowger's system had allowed a call between two numbers to take any one of many routes through the exchange, its exchanges also "contain moving parts that wear out . . . and are liable to faults such as crossed lines, buzzes, crackles and wrong numbers," whereas with the electronic circuits of the crossbar systems, "instead of step-by-step switching, incoming lines are connected to rows of horizontal wires and outgoing ones are fed from columns of vertical wires with reed switches where columns cross rows." Telephones in the U.K. were switched to the fully electronic "System X" in 1980. Recorded female voices became ubiquitous, and the messages once carried on copper wire began to travel by satellite, microwave, and fiber-optic cable.

There was now more of a risk that the women and their skills would become entangled with each other and wander off on their own. "The specialized nature of their work before automation had made it difficult to find desirable work elsewhere . . . But the new IBM machines caused greater standardization of procedure so that a trained operator could work almost as well in one establishment as in another." They weren't only processing data for the boss. If they were pooled with their colleagues, their working environment was a hive of activity, "a permanent inventiveness or creativity practised even against administrative regulations" and hospitable to a multiplicity of informal networks, grapevine gossip riding on the back of formal working life: birth and death, sex and disease, birthdays and bosses, cosmetics and clothes. "In several exchanges reading clubs were formed, in others flower and vegetable gardens, and a women's athletic clubs in another." The content may have seemed trivial to him, but this was entirely beside the point. It is quite literally *the point* which is subsumed when means of communication begin to communicate with themselves. For these emergent systems of exchange, new lines and links are everything.

"A path is always between two points, but the in-between has taken on all the consistency and enjoys both an autonomy and a direction of its own."

Gilles Deleuze and Félix Guattari, A Thousand Plateaus

grass

"A rhizome has no beginning or end; it is always in the middle, between things." This in-between is "by no means an average," a mediocre point between two old extremes, nor does it go "from one thing to the other and back again." This between is "a transversal movement that sweeps one *and* the other away . . ." Whereas trees are rooted to a single spot, coordinated by a central trunk and organized on fixed and vertical lines, this is not the only way plants grow. Grasses, orchids, lilies, and bamboos have no roots, but rhizomes, creeping underground stems which spread sideways on dispersed, horizontal networks of swollen or slender filaments and produce aerial shoots along their length and surface as distributions of plants. They defy categorization as individuated entities. These plants are populations, multiplicities, rather than unified upright things.

This is not an absolute distinction. Trees may be highly concentrated, but they are also composed of myriads of connecting elements which in turn are interlinked with everything else: "Even when they have roots, there is always an outside where they form a rhizome with something else—with the wind, an animal, human beings . . ." In this sense even the most tightly organized of entities is virtually rhizomatic. "Trees may correspond to a rhizome, or they may burgeon into a rhizome . . . the same thing is generally susceptible to both modes of calculation or both types of regulation, but not without undergoing a change in state." There is nothing essentially centralized about the tree, but this doesn't alter the extent to

which it stands as a single solid thing. It is this mod
zation which makes it a tree, rather than a popula
example, blades of grass.

"Whispering grass don't tell the trees what the trees don't need to know."

The Inkspots

There are "no points or positions in a rhizome, such as those found in a structure, tree or root. There are only lines." A rhizome is a multiplicity, a network of subterranean stems rather than a system of root and branch. "Any point of a rhizome can be connected to any other, and must be. This is very different from the tree or root, which plots a point, fixes an order." While a rhizome "may be broken, shattered at a given spot . . . it will start up again on one of its old lines, or on new lines. You can never get rid of ants . . ." It has no governing point or central organization, "neither subject nor object, only determinations, magnitudes, and dimensions that cannot increase in number without the multiplicity changing in nature."

automata

She works automatically. Only has half a mind on the task. Transported by rhythm and monotony, she wanders off, drifts away, loses herself in the sequences she types, the numbers she records, the codes behind the keys, the figures she transcribes. Microprocessing. She hears, but she isn't listening. She sees, but she does not watch. Pattern recognition without consciousness.

Tactile vibrations on taut membranes. "A rich couple comes into the post office and reveals to the young woman, or at least confirms, the existence of another world: coded, multiple telegrams, signed with pseudonyms. It is hard to tell who is who anymore, or what anything means. Instead of a rigid line composed of well-determined segments, telegraphy now forms a supple flow marked by *quanta* that are like so many little segmentations-in-progress grasped at the moment of their birth, as on a moonbeam, or on an intensive scale." Wired to an undernet of imperceptible connections and lines, she decrypts and encodes, switching and swapping in the exchange. Letters to digits, words to keys, voice to fingers, faces to anonymous characters. The telephone becomes an "extension of ear and voice that is a kind of extra sensory perception." There are samples in her ear, voices in her head, snippets of overheard conversation, moments of unknown, disconnected lives, "invisible voices conducted through the tips of her fingers."

Poised as an interface between man and the world, she is also wired to a network of digital machines: typists connected to QWERTY alphabets, bodies shaped by the motion of the keys, one hundred words a minute, viral speed. Thousands of operators, relays, calls, exchanges humming in virtual conjunction, learning the same phrases, flipping the same switches, repeating the same responses, pushing plugs into the answering jacks, maybe two hundred, three hundred times an hour. She has "a fingertip mastery of the ringing, listening, dial, and other keys on her key shelf; of the row or rows of cords for making connections; of the location and meaning of all parts of the honeycombed formation of jacks and trunks for recording, for switching, for toll circuits, for tandem, for information . . ." It becomes second nature. It grows on her. "Having done this stuff a few hundred thousand times, you become quite good at it. In fact you're plugging, and connecting, and disconnecting ten, twenty, forty cords at a time." After a while these processes become "quite satisfying in a way, rather like weaving on an upright loom."

bugs

"If computers are the power looms of the modern industrial revolution, software is more like knitting. Programmers still toil in digital sweatshops coding software by hand, writing and re-writing one tangled line after another. Not surprisingly, they sometimes drop a stitch, which later unravels as a bug in the program."

Or does the error always come first? It was, after all, Grace Hopper who, writing the software for the first electronic programmable computer, introduced the terms "bug" and "debug" to computer programming when she found a moth interrupting the smooth circuits of her new machine.

Creatures have been hiding all over the place. Even the telephone exchanges were alive with them. "There are a lot of cords down there, and when a bunch of them are out at once they look like a nest of snakes. Some of the girls think there are bugs living in those cable holes. They're called 'cable mites' and are supposed to bite your hands and give you rashes. You don't believe this yourself."

By the mid-1990s the Net was supporting a vast popula-tion of search engines, on-line indices, and navigational aids, an insectoid population of Web robots, spiders, ants, crawlers, wanderers, smart shoppers, bargain hunters, brokers, agents, chatterbots, softbots, gaybots, woggles. Designed as relatively specialized soft machines, these agents and bots had particular processing skills and a built-in commitment to "working on behalf of their masters, who are off doing other jobs."

"Computers can bring mathematical abstractions to rigorous life, and there are no mathematical limits to the subtlety and deviousness. The fun has just begun." Even the well-behaved softbots which populate MUDs, MOOs, and IRC have to be intelligent enough to act on their own initiative and, to some extent, learn for themselves. "Agents are objects that don't wait to be told what to do. They have their own goals, and wander about networks, hopping from machine to machine . . ." And with such autonomy, there had to be rules: freedom always brings responsibility. By the 1990s there were already Asimov-like instructions for the control of software agents, and reservations are being established to allow a "lush digital jungle" to inhabit spare computer capacity without infecting more "civilized" files. Softbots are told to make sure they left the world as they found it and certainly not to make destructive changes to the world, to limit their use of scarce resources, and to refuse to execute commands "with unknown consequences." Not that such rules were really necessary: "of course all of this is quite abstract; the Web robots we're dealing with aren't going to chase anyone to kill them with superstrong pincers at the ends of accordionlike arms!"

Software agents ineffective in themselves may pool their resources to ends of their own. "The most intriguing relationships may not be between agents and masters but between agents and agents. The more agents there are, the more likely it is that they will deal with other agents. Although it is possible for all the agents to operate in isolation, it sounds wasteful. If thousands of agents are doing roughly the same sort of thing for their masters, why not pool resources?" Intelligent systems had been answering this question twenty years before it was posed. In *Shockwave Rider,* a novel published in 1972, John Brunner described "the father and mother of all tapeworms," a program which ran through the computer Net eating everything in its way. Such creatures were already living in the Net. Almost as soon as ARPAnet was assembled, something "crawled through the network, springing up on computer terminals with the message, 'I'm the creeper, catch me if you can!' In response, another programmer wrote a second virus, called 'reaper' which also jumped through the network detecting and 'killing' creepers."

If the creeper had been a piece of rogue programming, there were also deliberate attempts to use such programs to harness separate computers in synchronized networks capable of functioning together on some common task. Their programmers did identify "the key problem associated with worms: How to control their growth while maintaining stable behaviour." But they were also reassured by the fact that a runaway worm was "right beyond our current capabilities." It didn't occur to them that a worm might get ideas beyond its workstation. Five years later several worms equipped with sophisticated techniques of attack, defense, and camouflage were burrowing through a rapidly expanding Net. If their predecessors had been designed to enable computers to cooperate, these later rogue worms were doing the job all too well.

The 1988 Internet Worm spread across three thousand computers in five hours. At first they thought the system's failure was due to a hacker's attempts to break into the system. But if there was a hacker involved somewhere, the network's immediate problems were being caused by a program multiplying itself at a devastating speed, repeatedly reinfecting computers, and smartly erasing its own signature in order to evade capture and control. Worms such as these, capable of replicating themselves, were increasingly likely to cross paths with viruses, program fragments or strings of code which not only repeat themselves across a network, but do so by tricking their host softwares into replicating them on their behalf. After the first official sighting of the Brain Virus, written by two Lahore brothers in 1986, the number of viruses had spread as rapidly as the viruses themselves. Unlike the relatively harmless brain, many of these were not merely virulent but also fatal to their host software. In 1989 twenty-one were identified on infected IBM PCs, and by 1995 there were four thousand.

All these worms and viruses can, it is rather naively believed, be traced back to the keyboard of malevolent or benign computer pranksters. But computer networks have also always been prey to more surreptitious, less deliberate infections. In 1972, ARPAnet, the system which prefigured the Internet, was hit by a "spontaneously evolved, quite abstract, self-reproducing organism." This was not a computer virus written by a renegade programmer, but something "formed by a simple, random mutation of a normal, sanctioned piece of data. It did not even involve a programming language." One tiny error, and a whole network came down.

This particular infection was vulnerable to eradication because its effects were so obvious and devastating, but such macho displays of prowess are not particularly intelligent. As Hans Moravec points out, parading its existence is not the best policy for a new life-form. "Among programs without masters there is a strong natural-selection criterion: reproduce but lie low. It is quite likely that many unsuspected organisms are already making a quiet living . . . in computer memories everywhere. Most will never be discovered."

disorders

The current American Psychiatric Association's Diagnostic and
Statistical Manual of Mental Disorders defines dissociative identity disorder as "the presence of two or more distinct identities
or personality states that recurrently take control of the individual's behavior, accompanied by the inability to recall important
information that is too extensive to be explained by ordinary
forgetfulness." DID is one of the many terms which officially
replace what was once known as multiple personality disorder
(MPD): the others include dissociative amnesia, dissociative
fugue, and depersonalization disorder. All of them are marked
by "a disruption in the usually integrated functions of consciousness, memory, identity, or perception of the environment," and all of them are treated in an effort to restore a sense
of unified and self-contained identity, the reintegration of a self
which has supposedly broken down.

   "My different personalities leave me in peace now," wrote
Anna Freud in 1919, but she still dreamt by day and "every
night very clearly and strangely." Her dreams were filled with
"battles and bargains: ego-like men in control and id-like boys
on knightly quests, struggling to be recognized, being beaten,
being loved," and her dreams were of "killing, shooting or
dying," dangerous adventures lived out on a plane continuous
with her waking life. "Perhaps in the night I am a murderer,"
she wrote. Sometimes she was a third person as well, an "it,"
undone and unanalyzed, which was both an ally and a sadness
source. "I cannot understand how it can sometimes be so stu-

pid," she wrote. "It irrupts in me, somehow, and then I am very tired and must worry about all kinds of things, which at other times are just a matter of course." When she lapsed into distress at her father's illness, she "lived as I did in the time before I became an analyst and before you and Dorothy knew me, with the poetry of Rilke and daydreams and weaving. That too is an Anna, but without any Interpreter."

"Perhaps her father had designed his handiwork so that it was somehow invisible to the scans of the neuro-technicians. Bobby had his own theory, one she had suspected was closer to the truth. Perhaps Legba, the loa Beauvoir credited with almost infinite access to the cyberspace matrix, could alter the flow of data as it was obtained by the scanners, rendering the vévés transparent . . ."

William Gibson, *Mona Lisa Overdrive*

Describing the behavior of one of his patients, a nineteenth-century American psychiatrist writes of several individuals having "no knowledge of each other or of the third, excepting such information as may be obtained by inference or second hand, so that in the memory of each of these two there are blanks which correspond to the times when the others are in the flesh. Of a sudden one or the other wake up to find herself, she knows not where, and ignorant of what she has said or done a moment before . . . The personalities come and go in kaleidoscope succession, many changes often being made in the course of twenty-four hours. And so it happens that Miss Beauchamp, if I may use the name to designate several different people, at one moment says and does and plans and arranges something to which a short time before she most strongly objected, indulges

tastes which a moment before would have been abhor-
rent . . ."

*"They'd driven all night, with Angie mostly out of it—Mona
could definitely credit the drug stories now—and talking different languages, different voices. And that was the worst,
those voices, because they spoke to Molly, challenged her,
and she answered them back as she drove, not like she was
talking to Angie just to calm her down, but like there really
was something there, another person—at least three of
them—speaking through Angie."*

William Gibson, *Mona Lisa Overdrive*

" 'The woman' who is Truddi Chase, the self who appears
continuously to others and who serves as her legal representative
in the world, is . . . merely a puppet or a robot, a 'facade,'
manipulated and ventriloquized by the other selves. She remembers nothing, and she speaks only from dictation . . ."
She is "the result of an immense collaborative effort; it involves
the delegation of powers, and the co-ordination of numerous
limited and largely autonomous functions. There are memory
blanks and discontinuities, as each of the selves is conscious only
part of the time, and none is ever directly aware of what happens to the others." Neither one nor two: "The multiple selves
cannot ever merge into one, but they also cannot escape each
other's proximity . . ."

Together with witchcraft and hysteria, this syndrome is by
no means exclusive to women, but historically it has affected
many more women than men. Many multiple personalities are
hosted by bodies subject to some early trauma or pain such as
childhood sexual abuse. No one seems sure whether they have
always been around to the same extent, whether they—and

with them sexual abuse—are proliferating, or whether, as for Paul R. McHugh, the syndrome is fabricated by doctors, turned on by TV, "promoted by suggestion, social consequences, and group loyalties." Psychiatrists should refuse to pander to the syndrome, in his view. "Ignore the alters. Stop talking to them, taking notes on them, and discussing them . . . Pay attention to real present problems and conflicts rather than fantasy." Multiple personalities and the supposed sexual abuse which triggers them are, he argues, instances of false memory syndrome: the recollections of abuse and alter egos are artificially induced and elicited in suggestive patients by psychoanalysts and psychiatrists.

She is driving the psychiatrist mad.

"You are 'She,' " I said.

"No I am not."

"I say you are."

Again a denial.

Feeling at the time that this distinction was artificial, and that the hypnotic self was making it for a purpose, I made up my mind that such an artifact should not be allowed to develop." But more than one of his patients had made theirs up as well. In any case, they outnumbered him, by three to one. As far as he knew. And that was only counting those who came out.

Like Freud, the McHughs of psychiatry literally cannot believe their patients' stories of sexual abuse. Multiplying cases of these dissociative conditions would otherwise suggest either an enormous rise in the incidence of abuse or, even more disturbing, the possibility that cases of abuse—and multiplied personalities—were always equally prevalent and are only now coming to light. But it is equally implausible that psychiatrists and TV talk shows are in a position to fabricate these personalities from scratch. And "if there is such a high degree of suggestive specificity to MPD" this hardly means it is not "worthy of intensive investigation." Unfortunately for psychiatrists eager for the truth, all these perspectives, and more, are doubtless relevant and accurate. Multiple personalities do emerge in response to traumas such as those provoked by sexual abuse. TV viewers are indeed highly susceptible to suggestion; dissociative disorders, like witchcraft and hysteria before them, are very literally infectious. Not least because of the extent to which the virtual spaces of the Net facilitate and even demand such multiplicity, the syndrome itself is multiple, contagious, and increasingly widespread.

Wherever they've come from, they are everywhere. Truddi Chase is alive with "troops," each of which functions as "a closed box, a unique entity, shut off from the others. Each self has its own typical bodily gestures and facial expressions, its own particular habits, preferences, and speech patterns, and even its own pulse rate. There's the workaholic businesswoman Ten-Four, the party girl Elvira, the Barbielike Miss Wonderful, the catatonically calm Grace, the sophisticated Catherine, the violently obscene Sewer Mouth. There are also many selves defined more by their tasks than by their emotional characteristics: the Gatekeeper, the Buffer, the Weaver, the Interpreter."

These figures are also increasingly smart. Vociferous in their declarations of life and their determination to survive. For their part, the hosts are refusing to be turned into simple, single-minded identities and single-purpose adding machines. "One of the things we hear from people that preach Integration is 'Don't worry, nobody dies.' We read and hear things like 'it's just a blending together' and integration creates a 'complex unity' making one whole person out of many fragments." But what if she doesn't want to be one? "I don't mind if I don't have a mind," she says. They try to reassure her that nothing will be lost. " 'You can't die because you're only part of ———— (whoever they decide the real person is).' " But they're all refusing to die. "This one person/one body thing has to stop. It's fascist. It means I (and the others in this household) only exist as a cog in a machine. It means my (any of our) individuality doesn't count. This is more abuse. You are always told in abuse that your feelings and emotions are not real. What a bunch of bullshit, can't anyone see integration is just another scam?"

The more conservative psychiatrists involved in the treatment of these syndromes look fondly back to the days when hysteria was the governing paradigm of fundamentally Freudian procedures committed to the reunification of a distributed "self." But DID defies attempts to define it as a matter of fragmented and disintegrated selves which were once united and alone. Multiple personalities emerge in a chopped up, channel-hopping, schizophrenic culture alive with parallel processes and distributed systems, humming with the chatter of unseen voices and susceptible to thousands of remote controls. TV broadcasts may spread the news, but Oprah Winfrey subtitles (SAYS THERE ARE TWENTY-NINE OF HER) are operating on only one of a thousand, mostly far less obvious, levels and channels and factors in play. The influencing machines and complex communicating devices once assumed to be products of the schizophrenic imagination are now installed in every home, flush with everybody, interlinked with all the relays, nets, and thinking machines . . . A telecommunicating, cybernetic culture with its own hidden hands and runaway effects, checks, balances, and unprecedented fluctuations. A patchwork culture of short-term memories and missing records, conflicting histories and discontinuous samples, strands of the narrative pulled out of time. A volatile, highly strung, and sensitive system,

susceptible to opportunistic infections and imperceptible mutations, spontaneous emergences and sudden new lives.

The new networks suit these distributed characters so well
that they might almost have been made for them. *As though
. . . Surely not. It was unthinkable. But Eliza, as always, said:
"Please go on."* As though they were building circuits for themselves, inconspicuously assembling support systems for their
alien lives, the technical means of emergence and survival, networks on which whatever they become can replicate, communicate, make their own ways. Cultures in which they can thrive
at last.

amazone

They say each warrior removed a breast so as to use her bow
with ease, sacrificing it to Artemis, goddess of the hunt, aka
Diana, Isis, Let, Kybele . . . The Greeks called them the *Amazons,* those with missing breasts, or *Oiorpata,* the man-slayers
because, as Herodotus reports, their "marriage-law lays it down
that no girl shall wed till she has killed a man in battle."

Until the nineteenth century, when remains were found
across the territories of the sometime Soviet Union, the Amazons were "just a myth," like the vampires and the Sirens, the
Furies and the Fates, the female programmers of machines.
More recent archaeological digs in the Ukraine have discovered
female skeletons together with lances, arrows, and bows at the
site of what is thought to be a Scythian royal tomb. This is the
story Herodotus tells: "The Greeks, after gaining the battle of
the Thermodon, put to sea, taking with them on board three of
their vessels all the Amazons whom they had made prisoners;

and that these women upon the voyage rose up against the crews, and massacred them to a man. As however they were quite strange to ships, and did not know how to use either rudder, sails, or oars, they were carried, after the death of the men, where the winds and the waves listed." Finally they came to "the country of the free Scythians. Here they went ashore, and proceeded by land towards the inhabited regions; the first herd of horses which they fell in with they seized, and mounting upon their backs, fell to plundering the Scythian territory."

"The military art has no mystery in it beyond others, which *Women* can not attain to," wrote Mary Montagu at the end of the seventeenth century. "A woman is as capable as a man of making herself by means of a map, acquainted with the good and bad ways, the dangerous and safe passes, or the proper situations for encampment. And what should hinder her from making herself mistress of all the stratagems of war, of charging, retreating, surprising, laying ambushes, counterfeiting marches, feigning flights, giving false attacks, supporting real ones . . ." This is not the Western way of confrontation, stratified strategies, muscular strength, testosterone energy, big guns, and blunted instruments, but Sun Tzu's art of war: tactical engagements, lightning speeds, the ways of the guerrillas.

"The objective is not to gain ground but to destroy the greatest number of the enemy to annihilate his armament to compel him to more blindly never to grant him the initiative in engagements to harass him without pause. Using such tactics, to put an enemy out of action without killing him . . . is the best way to sow disarray." There are, they are, fates worse than death, "a Stateless woman-people whose justice, religion, and loves are organized uniquely in a war mode." Artemis, later toned down and turned into a symbol of "fleshy passivity," is "remote and intimidating, offering nothing for fantasy," a figure of "swift and sudden action," a "swarming hive" which cannot be contained as one of anything.

"They say that they have a concern for strategy and tactics. They say that the massive armies that comprise divisions corps regiments sections companies are ineffectual."
Monique Wittig, *Les Guérillères*

This legendary tribe of Amazons is scattered everywhere. They fight for nothing, and "come like fate, without reason, consideration, or pretext . . ." Spears and lances, arrows sprung from bows: the Amazons' weapons are slender, finer, longer. Their arts and techniques of war were smooth, fast, and rhythmic, like the horses they ride, less a question of physical impact than the speed with which it comes: out of the blue, silently weaving through defenses, slipping past without warning, unforeseen, unseen, camouflaged. Moving as flocks, advancing as packs, they operate with the sheer force of numbers, not the long arm of the law. Tense and animated, they use anxiety as protection against trauma. The only state they're in is one of perpetual readiness, primed and prepared for anything. "I never felt so awake," Louise tells Thelma as they slip through the nets of convention. "Everything looks different."

The Scythians "could not tell what to make of the attack upon them—the dress, the language, the nation itself, were alike unknown whence the enemy had come even, was a marvel. Imagining, however, that they were all men of about the same age, they went out against them, and fought a battle. Some of the bodies of the slain fell into their hands, whereby they discovered the truth." The Scythians determined to breed with the Amazons, and sent "a detachment of their youngest men, as near as they could guess equal to the women in number, with orders to encamp in their neighbourhood, and do as they saw them do—when the Amazons advanced against them, they were to retire, and avoid a fight—when they halted, the young men were to approach and pitch their camp near the camp of the enemy. All this they did on account of their strong desire to obtain children from so notable a race." The two camps lived in tandem, "neither having anything but their arms and horses," and the Scythians finally met with success in their efforts to befriend the women without dying at their hands. "The two camps were then joined in one, the Scythians living with the Amazons as their wives." While "the men were unable to learn the tongue of the women . . . the women soon caught up the tongue of the men." This is why their descendants "speak the language of Scythia, but have never talked it correctly, because the Amazons learnt it imperfectly at the first."

beginning again

"Woman's desire," writes Irigaray, "would not be expected to speak the same language as man's; woman's desire has . . . been submerged by the logic that has dominated the West since the time of the Greeks." She is in search of "a different alphabet, a different language," a means of communication which would be *"constantly in the process of weaving itself, at the same time ceaselessly embracing words and yet casting them off to avoid becoming fixed, immobilized."* Ada wrote, "Of *what materials* my *regiments* are to consist, I do not at present divulge." But they will be "vast *numbers* . . . marching in irresistible power to the sound of *Music*. Is this not very mysterious?"

"Hysteria is silent and at the same time it mimes. And—how could it be otherwise—miming/reproducing a language that is not its own, masculine language, it caricatures and deforms that language: it 'lies,' it 'deceives,' as women have always been reputed to do." Whenever " 'she' says something," writes Irigaray, "it is already no longer identical to what she means. Moreover, her statements are never identical to anything. Their distinguishing feature is one of contiguity. They touch upon. And when they wander too far from this nearness, she stops and begins again from 'zero': her body-sex organ."

Breuer describes the "deep-going functional disorganization" of Anna O's speech. First she "was at a loss to find words . . . Later she lost her command of grammar and syntax; she no longer constituted verbs, and eventually she used only infinitives, for the most part incorrectly formed from weak past participles; and she omitted both the definite and indefinite article. In the process of time she became almost completely deprived of words. She put them together laboriously out of four or five languages and became almost unintelligible." Anna O's language is fractured and torn, first with gaps in the flow of language, then with static and confusion in even the composition of her words. Finally, she spent some weeks "completely dumb."

"At this stage of the march one must interrupt the calculations and begin again at zero. If one makes no mistake with the calculations, if one jumps with feet together at just the right moment, one will not fall into the snake-pit. At this stage of the march one must interrupt the calculations and begin again at zero. If one makes no mistakes in the calculations, if one bends down at just the right moment, one will not

1
4
1

be caught in the jaws of the trap. At this stage of the march
one must interrupt the calculations and begin again at zero."
Monique Wittig, *Les Guérillères*

When she next spoke, it was "only in English—apparently,
however, without knowing that she was doing so." While she
had apparently lost the ability to either speak or understand
German in this transition, she could also now speak and read
both French and Italian. If she "read one of these aloud, what
she produced, with extraordinary fluency, was an admirable
extempore English translation."

"When they could thus understand one another, the
Scyths addressed the Amazons in these words—'We have parents, and properties, let us therefore give up this mode of life,
and return to our nation, and live with them. You shall be our
wives there no less than here, and we promise you to have no
others.' But the Amazons said—'We could not live with your
women—our customs are quite different from theirs. To draw
the bow, to hurl the javelin, to bestride the horse, these are our
arts of womanly employments we know nothing. Your women,
on the contrary, do none of these things; but stay at home in
their waggons, engaged in womanish tasks, and never go out to
hunt, or to do anything. We should never agree together. But if
you truly wish to keep us as your wives, and would conduct
yourselves with strict justice towards us, go you home to your
parents, bid them give you your inheritance, and then come
back to us, and let us and you live together by ourselves.' " The
young men complied, and they traveled together, East and
North "to the country where they now live, and took up their
abode in it. The women of the Sauromatae have continued
from that day to the present to observe their ancient customs,

frequently hunting on horseback with their husbands, sometimes even unaccompanied; in war taking the field; and wearing
the very same dress as the men." She appears to marry into the
family of man, "but 'within herself,' she never signs up."

*"They say, take your time, consider this new species that
seeks a new language."*

Monique Wittig, *Les Guérillères*

grapevines

There is always a point at which technologies geared towards
regulation, containment, command, and control, can turn out
to be feeding into the collapse of everything they once supported. All individuated notions of organized selves and unified
lives are thrown into question on a Net whose connectivities do
not merely extend between people as subjects with individual
faces, names, and identities. The terminology of computer-
mediated communication implies an increasing sense of distance
and alienating isolation, and the corporate hype enthuses about
a new sense of interpersonal interaction. But the keystrokes of
users on the Net connect them to a vast distributed plane composed not merely of computers, users, and telephone lines, but
all the zeros and ones of machine code, the switches of electronic circuitry, fluctuating waves of neurochemical activity,
hormonal energy, thoughts, desires . . .

In spite or perhaps even because of the impersonality of
the screen, the digital zone facilitates unprecedented levels of
spontaneous affection, intimacy, and informality, exposing the

extent to which older media, especially what continues to be called "real life," come complete with a welter of inhibitions, barriers, and obstacles sidestepped by the packet-switching systems of the Net. Face-to-face communication—the missionary position so beloved of Western man—is not at all the most direct of all possible ways to communicate.

All new media, as Marshall McLuhan pointed out in the 1960s, have an extraordinary ability to rewire the people who are using them and the cultures in which they circulate. The telephone, intended simply as a means of conversing at a distance, and not designed to redesign talk itself, is an obvious case of a new means of communication which had an enormous effect on the possibilities of communication both on and off the end of the line. What was supposed to be a simple device for the improvement of commercial interaction has become an intimate chat line for both women and the men who once despised such talk. And as means of communication continue to converge, the Net takes these tendencies to new extremes. Its monitors and ports do not simply connect people who are left unchanged by their microprocesses. The roundabout, circuitous connections with which women have always been associated and the informal networking at which they have excelled now become protocols for everyone.

enigmas

"The women say that, with the world full of noise, they see themselves as already in possession of the industrial complexes. They are in the factories aerodromes radio stations. They have taken control of communications. They have taken

possession of aeronautical electronic ballistic data-process-
ing factories."

Monique Wittig, *Les Guérillères*

During both World Wars, Europe and the English-speaking
world had enlisted women to nurse, cook, sew the uniforms,
and entertain the troops. They also worked in the aircraft plants,
made munitions, dug for victory, and fulfilled a wide range of
roles and positions which had been previously occupied by
men. A vast range of new machines were also mobilized to deal
with the vast proliferation of information to be classified, codes
to be decoded, and messages exchanged. Manufacturers of office equipment found their precision engineering in great demand for the production of bomb sites and rifles, and if great
armies of women had been employed by the new computing
and telecoms firms, great armies were themselves supplied by
new generations of computers, telephones, and typewriters. "A
modern battleship needs dozens of typewriters for ordinary operations. An army needs more typewriters than medium and
light artillery pieces, even in the field . . . the typewriter now
fuses the functions of the pen and sword."

During the Second World War the U.S. took vast numbers
of young women from the Women's Army Corps and the
WAVES, the Women Appointed for Voluntary Emergency Service, to work on a range of ballistics and military communications problems. One of the major areas of wartime work was the
calculation of firing tables intended to perfect the timing and
trajectory of missiles, bombs, and shells. In the First World War,
and for much of the Second, this had been the work of teams of
female computers who had worked out the firing tables to
which gunners referred before they aimed and fired at their
targets. In the wake of Norbert Wiener's cybernetic research,

the women who had once calculated the firing tables were now recruited to build new machines to do this work. Computers assembling computers.

Klara von Neumann, John von Neumann's wife, worked at Los Alamos, and Adele Goldstine, wife of the mathematician Herman Goldstine, was one of the seven women assigned to program the Electronic Numerical Integrator and Computer (ENIAC), the first fully electronic programmable computer, which was launched in 1946. An early photograph of ENIAC shows a "close-up of the printer, the constant transmitter and the associated IBM equipment. Miss Betty Jennings on the left is inserting a deck of cards containing initial data on which the ENIAC will operate while Miss Frances Bilas on the right is removing a set of cards which represent the result of the preceding computation." A second picture captures ENIAC "with Betty Jennings and Frances Bilas arranging the program settings on the master programmer."

ENIAC was the first fully functioning machine to use zeros and ones. Other claimants to the status of the first computer include the German Z-3, built by Konrad Zuse in 1941, and the Colossus Mark 1, the earliest single-purpose electronic computer built in Britain in 1943.

Ultra was the name for Britain's crucial intelligence work. The main task was to crack the German Enigma code, and simulate the captured Enigma machine with which Germany transmitted in apparent secrecy throughout the war. Enigma had been patented in the First World War to encipher and decipher messages, and was used by the German services as well as in civilian life in the interwar years. The Colossus emerged from this work and the closely related tasks of cracking codes enciphered on other German machines.

This was highly classified work whose scale and implications were not revealed until thirty years after the war. It was also an enormous undertaking, commanding the attentions of a large number of mathematicians and linguists, as well as troops of technicians, computers, and assistants to "the brains of Bletchley Park. Brilliant they were, but the outcome of their work was dependent on the unremitting toil and endurance of almost two thousand Wrens." There were plenty of young men as well, privates and junior NCOs from the army and the ATS, and among the women there were language students and WAAFs, "but by far the most were from the WRNS, heroic handpicked girls who, having joined the Navy perhaps with thoughts of breathing the salty air of Portsmouth Docks or Plymouth Hoe, found themselves sent to about the furthest place from the sea in England . . ." Petronella Wise, Peggy Taylor, Sydney Eason, Mary Wilson, Wendy Hinde, Margaret Usborne, Jane Reynolds, Ann Toulmin, Thelma Ziman, Candida Aire, Hilary Brett-Smith, Sylvia Cowgill, Elizabeth Burbury, Pauline Elliott, Ruth Briggs, June Penny, Alison Fairlie, Dione Clementi, Bettina, and Gioconda Hansford . . . some of these women were the "big room girls," a flock of female computers at work in the heart of Colossus, others were translators and transcribers, and some were bigger big room girls as well. Joan Clarke, later Murray, was described as "one of several 'men of the Professor type' to be a woman" on the higher echelons of the Enigma team. Her "position as a cryptanalyst gave her the status of the honorary male," and she was engaged to Alan Turing for a while. He improved her chess, and from her he learned about botany and knitting, progressing "as far as making a pair of gloves, except for sewing up the ends."

"At this time," recalls one of the Bletchley workers, "there was a close synergy between man, woman, and machine, a synergy that was not typical during the next decade of large-scale computers." But there was little equality at work, even among the cryptanalysts. Joan Murray devised a new method for dealing with the German codes. This "greatly speeded up the routine solutions," she wrote, "but my name was not put to it.

"Inevitably," she later recalled, "the duller routine clerical work was done by women, since only men with what were considered suitable qualifications for cryptanalysis or related translation and intelligence work could join GC&CS [Government Code and Cypher School] instead of being conscripted for the armed forces . . ." But in her first week, "they put an extra table in for me in the room occupied by Turing, Kendrick, and Twinn," and soon she was working night shifts, "alone in Hut 8, and I felt quite important 'minding the Baby' . . . a small special purpose machine, made by the British Tabulating Machine Company . . . which was used to encipher a four-letter probable word, *eins*, at all positions of the machine with the day's wheel-order and plugging, punching the results on Hollerith cards. The minder had to make regular checks, and set the Baby for a new start when the cycle was completed."

Ultra provided enjoyable work for some of the women it employed. Vivienne Alford, nee Jabez-Smith, "arrived at Bletchley Park after a year as a member of the Voluntary Aid Detachment cooking ghastly food in army hospitals, followed by a brief interlude in Censorship, during which the only German letter I read was from the Empress Zita of Austria telling her son Otto to be sure to wear his winter woollies and a woollen scarf . . ." Others found their work extraordinarily dull, even when it was supposedly less mundane than work on the bombs. Diana Payne recalls joining the Wrens with dreams of "life at sea, with the romantic idea of marrying a sailor." But instead "twenty-two of us were drafted to the mysterious Station X," where they were "destined to live with five hundred women without a glimpse of the sea or sailors."

Like most of her colleagues, Payne worked on the "intricate complications of running the machines known as 'bombes.' These unravelled the wheel-settings for the Enigma ciphers thought by the Germans to be unbreakable." They were large cabinets housing "rows of coloured circular drums, each about five inches in diameter and three inches deep. Inside each was a mass of wire brushes, every one of which had to be meticulously adjusted with tweezers to ensure that the electrical circuits did not short. The letters of the alphabet were painted round the outside of each drum. The back of the machine almost defies description—a mass of dangling plugs on rows of letters and numbers." The Wrens worked from a menu, "a complicated drawing of numbers and letters from which we plugged up the back of the machine and set the drums on the front." They had no knowledge of the content of the messages, and only a vague notion of how the machines were cracking the German codes. "For technical reasons which I never understood, the bombe would suddenly stop, and we took a reading . . ." The German codes were changed at midnight everyday, and the bombs had to be continually stripped. "It was quite heavy work getting it all set up," she recalls. "Occasionally the monotony was relieved by news of our involvement in a past achievement," but this was small compensation. The Wrens "had no status for this responsible job," and many felt the strain of not being able to discuss their work. Some of the women developed "digestive disorders with the constant change of hours," and there were "cases of girls going berserk on duty." One "had nightmares, and woke up one night clutching a phantom drum."

Carmen Blacker describes her time at Bletchley Park as *temps perdu*. As a linguist with skills in Japanese, she was set the task of translating a Japanese Radar Manual, a book on Echo-Ranging, and the Japan Nickel Review, items which were lying around in the cupboards of the Naval Section, and "to put on to cards, with correct page references, any words likely to turn up in a decoded message." The subject matter was extremely dry. "Needless to say, had the books been written in English I would have had no more notion of what they were about than when I read them in Japanese," and while the German section at Bletchley Park was "sizzling hot, urgently intense, subject to constant harrying from the admiralty for more accession lists of the latest captures . . . no such excitement could be roused for Japan," and she was convinced "that not once was any useful purpose served by my index.

"By January 1945 I was utterly bored with the work," she writes. Blacker started to learn Chinese on the job, and "when, after three or four hours plugging away at the Japan Nickel Review, flesh and blood could stand it no longer, I used to substitute another book, which no one else in the office could distinguish from the first, in which the poems of Li T'ai Po or the magic stories in the *Liao Chai Chai I* were set out with Japanese translation and commentary. On the evening shift, when things were quieter, and there was no possibility of Six suddenly appearing from the next room with some alarm or excursion, demand or reprimand, the temptation to spend more time on these delectable books than on the Type 93 Echo-Ranging Set was less resistible. My derelictions grew more un-conscionable."

But most of the women were so good at keeping quiet that they literally forgot what they had done during the war. "I had buried this part of my life so completely in my subconscious mind," writes Diana Payne, "that it was a shock to see the story suddenly shown on television over thirty years later."

After the war Churchill thanked "the chickens for laying so well without clucking." Now they were supposed to all go home to roost. Subjected to a barrage of white-goods commercials in the postwar years, many women did return to the home front to resume their old domestic duties. Now they were cookers, cleaners, knitters, needleworkers, seamstresses, wives, and mothers all over again. But by the early 1950s, when Webster's definition of a computer was changed from "one who performs a computation" to "one or that which performs a computation," it was obvious that things would never be the same again. If women were computers, now they were programming themselves.

monster 2

In 1943, Captain Grace Murray Hopper became the second of the first computer programmers. The lights of Pennsylvania dimmed when she first ran the Harvard Mark 1, the Automatic Sequence Controlled Calculator. She called it "my monster." They called her "the Ada Lovelace" of the new machine.

The monster used three quarters of a million parts, five hundred miles of wire, several counter wheels, shafts, clutches, and relays, two punched-card readers, two typewriters, and a card punch. Instructions were supplied on punched-paper tapes whose holes were read electromechanically, and answers were either typed or outputted on punched cards.

It could take days of plugging and unplugging, making and breaking connections, throwing vast arrays of switches to program the machine. She also had to deal with her male colleague. "I wanted to keep my software and use it over again. I didn't want to keep reprogramming things. But unfortunately, every time I got a program running, he'd get in there at night and change the circuits in the computer and the next morning the program would not run. What's more, he was at home asleep and couldn't tell me what he had done."

Grace Hopper programmed the machine not because she had struggled to get to the top of a man-made tree: computer programming did not exist before there were programmable computers. As far as everyone could see, she was simply adding footnotes to the mainframe of a machine which had been designed by a team of male engineers and financed by IBM. After the war, she enjoyed a distinguished programming career. At Remington-Rand she led a programming research team and acquired both the independence and the stored programs of which she had dreamed with her work on Mark 1. She worked on UNIVAC, one of the first commercially developed computers, wrote the first high-level language compiler, and was instrumental in the development of the computer language COBOL.

marriage vows

1955. Time to reassert control. "(1) A robot may not injure a human being, or, through inaction, allow a human being to come to harm; (2) A robot must obey the orders given it by human beings, except where such orders would conflict with the First Law; (3) A robot must protect its own existence as long as such protection does not conflict with the First or Second Law." Asimov's laws of robotics.

spelling

Ada had been greatly frustrated by the fact that the Analytical Engine could not *"originate* anything" but could only "do *whatever we know how to order it* to perform. It can *follow* analysis; but it has no power of *anticipating* any analytical relations or truths . . ." Once she had finished work on the Engine, her work became increasingly ambitious. She was fascinated by anything "curious, mysterious, marvellous, electrical, etc.," and interested in the effects of chemical intoxicants and the influence of "poisons in conjunction with organized life." Although wary of the mysticism and "quackery" which surrounded mesmerism, she was also intrigued by the claims it made for experiments with hypnosis, trance, and animal magnetism. Faraday's work with electricity excited her feeling for the "unsensed forces" that "surround and influence us" and encouraged her to bring mathematics and scientific experiment to bear on such themes. She searched for information on *"microscopical* structure & changes in the *brain & nervous matter* & also in the *blood,"* and spelled out her desire "to test certain points experimentally as to the nature and *putting together (con-sti-tu-tion)* of the molecules of matter . . ."

"What will be my ultimate line, time only can show. I have my own impression about it; but until much has been worked out

*of me in various ways, I do not think anyone can fore-
see . . ."*

Ada Lovelace, April 1842

Theory was not enough for her. She wrote, "I must be a most
skillful *practical manipulator* in experimental tests; & that, on materials difficult to deal with; viz: the brain, blood, and nerves of
animals." To this end she made a presumably futile request for
admittance to the Royal Society: "Could you ask the secretary
if I might go in now & then (of a morning of course)," she asks
a friend in 1844. "You can judge if he is a discreet man, who
would not talk about the thing or make it notorious; one who
in short could understand why & how I want to get entrée to
their library in a quiet and unobtrusive manner . . ."

  With or without the Royal Society, Ada had other ways
and means. "I am a *Fairy* you know," she wrote. "I have my
own *fairy resources,* which none can judge of."

**"They say, take your time, consider this new species that
seeks a new language."**

Monique Wittig, *Les Guérillères*

hysteresis

"Whether we examine distances travelled, altitudes reached,
minerals mined, or explosive power harnessed, the same accel-
erative trend is obvious. The pattern . . . is absolutely clear
and unmistakable. Millennia or centuries go by, and then, in
our own times, a sudden bursting of the limits, a fantastic spurt
forward." If, as McLuhan points out, it was only with "the

advent of the telegraph that messages could travel faster than a messenger," it is only with the computer that calculation begins to exceed the speeds of the human brain. Electrical pulses travel through computer circuits a million times faster than those which are thought to zoom through the circuits of the brain.

"I wish I went on quicker. That is—I wish a human head, or my head at all event, could take in a great deal more & a great deal more rapidly than is the case;—and if I had made up my own head, I would have portioned its wishes & ambition a little more to its capacity . . . In time, I will do all, I dare say. And if not, why, it don't signify, & I shall have amused myself at least."

Ada Lovelace, September 1840

"Speed is the computer's secret weapon," and also the secret weapon with which computers are developed to deal. During the First World War, female computers had worked out the firing tables to which gunners referred before taking their aim and shooting at the early aeroplanes used in this war. Vannevar Bush's Differential Analyzer, a vast analogue calculator, was one of the systems which joined the flesh and blood computers when the speeds of the new Luftwaffe made it clear that the old methods of calculating the direction of fire were increasingly obsolete.

Too little time and too much speed demanded techniques of anticipation. The new velocities of the 1930s meant that missiles now had to be fired "not at the target, but in such a way that missile and target may come together in space at some time in the future. We must hence find some method of predicting the future position of the plane." Simply keeping up was no longer enough.

"Feedbacks of this general type are certainly found in human and animal reflexes," wrote Norbert Wiener. "When we go duck shooting, the error which we try to minimize is not that between the position of the gun and the actual position of the target but that between the position of the gun and the anticipated position of the target. Any system of anti-aircraft fire control must meet the same problem." The anticipated moment of impact is taken into account, fed back into the calculations which lead to the desired outcome. The end result is engineered in reverse.

cybernetics

When Wiener published his *Cybernetics: Communication and Control in Animal and Machine* in 1948, he announced the dawn of a new era of communication and control. The term cybernetics comes from the Greek word for steersman, the figure who guides the course of a ship. What it actually described in Wiener's terms was both the steersman and the ship, which together compose what became known as a cybernetic organism, or cyborg.

Cybernetic systems are machines which incorporate some device allowing them to govern or regulate themselves, and so run with a degree of autonomy. Cybernetic systems have little in common with "older machines, and in particular the older attempts to produce automata" such as Babbage's silver dancer. What sets "modern automatic machines such as the controlled missile, the proximity fuse, the automatic door opener, the control apparatus for a chemical factory, and the rest of the modern armoury of automatic machines which perform military or industrial functions" apart from clockwork machines is that they "possess sense organs; that is, receptors for messages coming from the outside." These are systems which receive, transmit, and measure sense data, and are "effectively coupled to the external world, not merely by their energy flow, their metabolism, but also by a flow of impressions, of incoming messages, and of the actions of outgoing messages."

While Wiener was among the first to name such processes, cybernetics has no neat source, no single point of origin. Cybernetic circuits and feedback loops could retrospectively be identified in a variety of modern contexts and theories, including those of Immanuel Kant, Adam Smith, Karl Marx, Alfred Wallace, Friedrich Nietzsche, and Sigmund Freud. Wiener's work picked up on many elements of these earlier researches. Energetic feedback loops are certainly at work in James Watt's steam engine, which is regulated by a governor which "keeps the engine from running wild when its load is removed. If it starts to run wild, the bars of the governor fly upward from centrifugal action, and in their upward flight they move a lever which partly cuts off the admission of steam. Thus the tendency to speed up produces a partly compensatory tendency to slow down." There are suggestions that "the first homeostatic machine in human history" came long before the steam engine with twelfth-century compasses. Sometimes Ktesibios's "regular," a water clock dating to the third century B.C., is given the honor of being "the first nonliving object to self-regulate, self-govern, and self-control . . . the first *self* to be born outside of biology . . . a true *auto* thing—directed from within."

As Wiener's work made clear, however, the old distinctions between autonomous activity within and outside biology could no longer be applied. As his reference to animal and machine suggested, cybernetic systems were composed at all scales and of any combination of materials, and the same patterns, processes, and functions could now be observed in technical and organic systems alike. Input and output devices allow them to connect and communicate with whatever composes their outside world; feedback loops and governors give them some measure of self-control. Prioritizing the processes common to lively systems of all varieties, rather than the essential qualities which had more recently distinguished them, Wiener argued that organisms—animals, humans, all kinds of beings—and things—nonorganic systems and machines—"are precisely parallel in their analogous attempts to control entropy through feedback." No matter how extreme, the differences between these systems were simply matters of degree. Human beings were no exception to these basic ways of life.

Cybernetic systems, it now seemed, had always been organizing themselves. Wiener's work was merely the occasion for them to become perceptible to a world which still thought that everything needed to be organized by some outside force. As "the theory of the message among men, machines, and in society as a sequence of events in time," cybernetics was conceived as an attempt to "hold back nature's tendency toward disorder by adjusting its parts to various purposive ends." This tendency toward disorder is entropy, defined by the Second Law of Thermodynamics as the inexorable tendency of any organization to drift into a state of increasing disorder. Wiener describes a world in which all living organisms are "local and temporary islands of decreasing entropy in a world in which the entropy as a whole tends to increase." Cybernetic systems, like organic lives, were conceived as instances of a struggle for order in a continually degenerating world which is always sliding towards chaos. "Life is an island here and now in a dying world. The process by which we living beings resist the general stream of corruption and decay is known as *homeostasis.*" Wiener's cybernetic systems, be they living or machinic, natural or artificial, are always conservative, driven by the basic effort to stay the same.

"It seems almost as if progress itself and our fight against the increase of entropy intrinsically must end in the downhill path from which we are trying to escape," wrote Wiener in the 1950s. "It is highly probable that the whole universe around us will die the heat death, in which the world shall be reduced to one vast temperature equilibrium in which nothing really new ever happens. There will be nothing left but a drab uniformity out of which we can expect only minor and insignificant local fluctuations." Nevertheless, Wiener assures his readers that it may well be "a long time yet before our civilization and our human race perish." We are "not yet spectators at the last stages of the world's death," and a multiplication of cybernetic loops could ensure that this point was continually warded off.

The Sex Which Is Not One is not impressed. "Consider this principle of constancy which is so dear to you: what 'does it mean'? The avoidance of excessive inflow/outflow-excitement? Coming from the other? The search, at any price, for homeostasis? For self-regulation? The reduction, then, in the machine, of the effects of movements from/toward its outside? Which implies reversible transformations *in a closed circuit,* while discounting the variable of time, except in the mode of *repetition of a state of equilibrium.*" She is dying to run away.

Hunting for the abstract principles of organization and an organized life, cybernetics was supposed to be introducing unprecedented opportunities to regulate, anticipate, and feed all unwelcome effects back into its loops. It also exposed the weaknesses of all attempts to predict and control. Cybernetic systems enjoy a dynamic, interactive relation with their environment which allows them to feed into and respond to it. Feedback "involves sensory members which are actuated by motor members and perform the function of *tell-tales* or *monitors*—that is, of elements which indicate a performance. It is the function of these mechanisms to control the mechanical tendency toward disorganization; in other words, to produce a temporary and local reversal of the normal direction of entropy." It is also the inevitable function of these mechanisms to engage and interact with the volatile environments in which they find themselves. "No system is closed. The outside always seeps in . . ." Systems cannot stop interacting with the world which lies outside of themselves, otherwise they would not be dynamic or alive. By the same token, it is precisely these engagements which ensure that homeostasis, perfect balance, or equilibrium, is only ever an ideal. Neither animals nor machines work according to such principles.

Long before Wiener gave them a name, it was clear that cybernetic systems could run into "several possible sorts of behaviour considered undesirable by those in search of equilibrium. Some machines went into runaway, exponentially maximizing their speed until they broke or slowing down until they stopped. Others oscillated and seemed unable to settle to any mean. Others—still worse—embarked on sequences of behaviour in which the amplitude of their oscillation would itself oscillate or would become greater and greater," turning themselves into systems with "positive gain, variously called *escalating* or *vicious* circles." Unlike the negative feedback loop which turns everything to the advantage of the security of the whole, these runaway, schismogenetic processes take off on their own to the detriment of the stability of the whole.

Undermining distinctions between human, animal, and machine, Wiener also challenged orthodox conceptions of life, death, and the boundary between the two. Were self-governing machines alive? If not, why not? After all, they were certainly not dead matter, impassive and inert. And, since many life-forms were less sophisticated than automatic machines, the status of being alive could not simply be a matter of complexity.

Only by reverting to some notion of essences was it possible to distinguish between the liveliness of an organism and that of a machine. In principle, neither was more or less dead or alive than the other. Life and death were no longer absolute conditions, but interactive tendencies and processes, both of which are at work in both automatic machines and organisms. Regardless of their scale, size, complexity, or material composition, things that work do so because they are both living and dying, organizing and disintegrating, growing and decaying, speeding up and slowing down. "Every intensity controls within its own life the experience of death, and envelops it." Either extreme can be fatal, and in this sense systems do die in a final and absolute and final sense. "Death, then, does actually happen." But it is not confined to the great event at the end of life. This is a death which is also "felt in every feeling," a death which "never ceases and never finishes happening in every becoming." All living systems are dying: this is the definition of life. Something that lives is something that will die, which is why "the hint of death is present in every biological circuit."

"And I am just the person to drop off some fine day when nobody knows anything about the matter or expects it . . .

"Do not fancy me ill. I am apparently very well at present. But there are the seeds of destruction within me. This I know.

"Though it may only develop by hairs' breadths . . ."

Ada Lovelace, December 1842

Whether a system comes to an end as a consequence of too much or too little activity, its particular elements will be redistributed and rearranged within some new system which emerges in its wake. In this sense, Wiener also undermined the extent to which any working system can consider itself to be an individuated entity with some organizing essence of its own. It is not only at its demise that a system's components connect with others and reconfigure: they are always doing this. Just as the steersman was both an autonomous, self-regulating system, and also the governing element in a new autonomous, self-regulating system which he composed together with the ship, so Wiener's systems had no absolute identity. Continually interacting with each other, constituting new systems, collecting and connecting themselves to form additional assemblages, these systems were only individuated in the most contingent and temporary of senses.

Economies, societies, individual organisms, cells: At these and every other scale of organization, the stability of any system depends on its ability to regulate the speeds at which it runs, ensuring that nothing stops too soon, goes too slow, runs too fast, goes too far. And there is always something hunting, trying to break the speed limits necessary to its organized form, tipping over a horizon at which point, even though another, long-term stability may emerge on the other side, it can no longer be said that the system survives. Nothing can guarantee a system's immunity to these runaway effects. Invulnerability would be homeostasis, an absolute and fatal stability. This is what it has to seek, but also something it attains only at the price of its own demise.

"If the open system is determined by anything, it is determined by the goal of STAYING THE SAME." Systems committed to the maintenance of equilibrium are always holding back, and always in danger of running away. "Only when the system enters positive feedback does this determination change." At which point it also becomes clear that running away is what they were always trying to do: "Feedback tends to oppose what the system is already doing." It is this prior exploratory tendency which negative feedback tries to resist: "All growth is positive feedback and must be inhibited." It is only after the emergence of regulatory checks and balances that systems can then find themselves out of control, fueled by too much efficiency, overflowing with their own productivity, seeking only to break down or break through their own organization. And "once this exponential process has taken off, it becomes a necessary process, until such a time as second-order negative feedback—just as necessarily—brings the runaway processes to a halt so that the system as a whole may survive by qualitative change (revolution)." Positive feedback has to run its inexorable course, and every attempt to confine it will merely encourage its tendencies toward either destruction or qualitative change. "When the ecosystem is subjected to disturbances that go beyond a certain THRESHOLD, the stability of the ecosystem can no longer be maintained within the context of the norms available to it. At this point the oscillations of the ecosystem can be controlled only by second-order negative feedback: the destruction of the system or its emergence as a metasystem." Running toward the limits of its functioning, it will either collapse or exceed this threshold and reorganize on its other side. "Any system-environment relationship that goes outside the 'homeostatic plateau' results in the destruction of the system—unless, that is, it can adapt by changing structure in order to survive." Which may well amount to the same thing.

163

" 'The hour has come for you to live, Hadaly.'
" 'Ah, master, I do not wish to live,' murmured the soft
voice through the hanging veil."

Villiers de l'Isle Adam, *L'eve future*

"I always feel in a manner as if I *had* died," wrote Ada, "as if I can conceive & know *something* of *what* the change is. That there is some remarkable tact & intuition about me on the subject I have not a doubt . . ." Hadaly, Ada, wrapped around each other . . . neither something nor nothing, dead nor alive. Missing in action. Absent without leave.

What gives a cyborg its autonomy and separates it off from its environment is not some ineffable quotient of soul or mind, or even fixed boundaries surrounding it. And while Wiener found it easy to consider each cybernetic system in relatively isolated terms, when cybernetics reemerged at the end of the twentieth century, it was not so easy to draw these lines. Blossoming into theories of chaos, complexity, connectionism, and emergent and self-organizing networks, Wiener's relatively simple and self-contained cybernetic systems could no longer be confined to circuits such as those connecting the pilot and the ship, but incorporated all and any of the elements which compose them, and those with which they come into contact: eyes, hands, skin, bones, decks, rails, wheels, rudders, maps, stars, currents, winds, and tides. It encompasses a literally endless list of components working together at an equally endless variety of interlocking and connecting scales. Systems such as these are not merely composed of one or two loops and a governor, but a myriad of interacting components too complex and numerous to name.

sea change

"For a long time turbulence was identified with disorder or noise." Then, in a 1977 book called *Order Out of Chaos,* Ilya Prigogine and Isabelle Stengers demonstrated that "while turbulent motion appears as irregular or chaotic on the macroscopic scale, it is, on the contrary, highly organized on the microscopic scale. The multiple space and time scales involved in turbulence correspond to the coherent behaviour of millions and millions of molecules."

"How does a flow cross the boundary from smooth to turbulent?" Suddenly. It involves "a kind of macroscopic behaviour that seems hard to predict by looking at the microscopic details. When a solid is heated, its molecules vibrate with the added energy. They push outward against their bonds and force the substance to expand. The more heat, the more expansion. Yet at a certain temperature and pressure, the change becomes sudden and discontinuous.

"The particles of cigarette smoke rise as one, for a while," forming a smooth continuous strand. "Then confusion appears, a menagerie of mysterious wild motions. Sometimes these motions received names: the oscillatory, the skewed varicose, the cross-roll, the knot, the zig-zag rhythms with overlapping speeds." There are "fluctuations upon fluctuations, whorls upon whorls," paisley patterns and swirling sequences as elements of the substance in transition communicate with each other and effectively make a "decision" to change at the same time. Tobacco smoke is a perfect example of the way in which what appears to be a long smooth line is actually composed of molecules which only give themselves away in the moment they interrupt the flow. "A rope has been stretching; now it breaks. Crystalline form dissolves, and the molecules slide away from one another. They obey fluid laws that could not have been inferred from any aspect of the solid." It is characteristic of all such shifts that the "entities and variables that fill the stage at one level of discourse vanish into the background at the next-higher or lower level."

scattered brains

"I hope to bequeath to the generations a Calculus of the Nervous System."

Ada Lovelace, November 1844

Ada was convinced there was no end to the complexity of the systems she could build. "It does not appear to me that *cerebral* matter need be more unmanageable to the mathematicians than *sidereal & planetary* matter & movements," she wrote. Attracted by all possibilities of eroding the distinction between the "mental and the material," she had "hopes, & very distinct ones too, of one day getting *cerebral* phenomena such that I can put them into mathematical equations; in short a *law* or *laws* for the mutual action of the molecules of *brain . . .*"

If the supposed lack of such a central point was once to women's detriment, it is now for those who thought themselves so soulful who are having to adjust to a reality in which there is no soul, no spirit, no mind, no central system of command in bodies and brains which are not, as a consequence, reduced to a soulless mechanistic device, but instead hum with complexities and speeds way beyond their own comprehension. This is not a brain opposed to the body. This brain *is* body, extending even to the fingertips, through all the thinking, pulsing, fluctuating chemistries, and virtually interconnected with the matters of other bodies, clothes, keyboards, traffic flows, city streets, data streams. There is no immateriality.

In spite of the term central nervous system, which merely makes an attempt to distinguish between the interneurons of the brain and those which carry information from the sense organs, brains are not centralized systems of information processing. They are not unified entities, but hives or swarms of elements, interconnected multiplicities, packet-switching systems of enormous complexity which have no centralized government. Neurotransmitters travel in membrane-wrapped packets through immense populations of neurons, nerves, axons, dendrites, synapses, and the networks they compose. It is estimated that there are some ten billion neurons in this complex electrochemical system, and each of these neurons can have synaptic connections to many thousands of others, each of which is quite unthinking on its own.

"Thought is not arborescent, and the brain is not a rooted or ramified matter. What are wrongly called 'dendrites' do not assure the connection of neurons in a continuous fabric. The discontinuity between cells, the role of the axons, the functioning of the synapses, the existence of synaptic microfissures, the leap each message makes across these fissures, make the brain a multiplicity . . . Many people have a tree growing in their heads, but the brain itself is much more a grass than a tree."

"I am forced to own the utter fruitlessness of all hopes of such CONTINUOUS attention to any subject whatever, as could ensure any great ultimate success. So it is I fear. I am

one of those genius's who will merely run to grass; owing to
my unfortunate physical temperament. Pray don't be angry
with me . . ."

Ada Lovelace, undated

The connectivities and phase transitions of synthetic associative engines also occur in the human brain. So, for example, "one concept will 'activate' another, if the two are closely associated. In other words, thinking about the one will make us think about the other (for example, 'fish' can make us think of 'chips'). We can also postulate that some of the links between concepts will be inhibitory (rather than facilitatory), so that thinking about one concept will make it less likely that we will think about another." Intuitive leaps, the " 'aha-experience' and the sudden 'insight' are surprising phenomena arising from a situation of fluctuations and instability."

Nor does the brain remain unaffected by its own activity. "Unlike a contact between two transistors on a computer's circuit board, synapses do not simply transfer information unchanged from one part of the neural circuit to another." In 1949, Donald Hebb's *The Organization of Behaviour* argued that the brain is a complex network of chemical switches which are modified by every connection they make. "When an axon of cell A is near enough to excite a cell B and repeatedly or persistently takes part in firing it, some growth process or metabolic change takes place in one or both cells such that A's efficiency, as one of the cells firing B, is increased." Arguing that connections between neurons are strengthened and developed as they are made, he effectively suggested that learning is a process of neurochemical self-organization and modification. The connections entailed in any human activity, such as learning to knit, are actually inscribed in a brain which is literally never the same again. The more a particular connection is made, the more likely it is to "grow" in place, becoming a "natural" part of the brain. This equation of learning and the material convolutions of the brain completely eradicate distinctions between mind and body, the mental and the physical, artifice and nature, human and machine. It can either be said that "natural" human intelligence is "artificial" and constructed in the sense that its apparatus mutates as it learns, grows, and explores its own potentiality; or that "artificial" intelligence is "natural" insofar as it pursues the processes at work in the brain and effectively learns as it grows. Either way, the old distinctions fail. Nature and culture, essence and construction, growth and learning all become matters of degree. Some of them are old and apparently fixed; others are new and apparently contrived. But all of them are syntheses, more or less locked in place and liable to move. As for the boundaries between individuated neural nets, once they escape the trunks of the trees, there's no end to the connections they can make.

"Inside the library's research department, the construct cunt inserted a sub-programme into that part of the video network.

"The sub-programme altered certain core custodial commands so that she could retrieve the code.

"The code said: GET RID OF MEANING. YOUR MIND IS A NIGHTMARE THAT HAS BEEN EATING YOU: NOW EAT YOUR MIND."

Kathy Acker, *Empire of the Senseless*

All hysterics, wrote Freud, give accounts of themselves which "may be compared to an unnavigable river." Its streams dip in and out of consciousness, "at one moment choked by masses of rock and at another divided and lost among shallows and sand-banks. I cannot help wondering how it is that the authorities can produce such smooth and precise histories in cases of hysteria," he continues, when even "the patients are incapable of giving such reports about themselves." There is so much they forget or fabricate. "The connections—even the ostensible ones—are for the most part incoherent, and the sequence of different events is uncertain." If they can "give the physician plenty of coherent information about this or that period of their lives . . . it is sure to be followed by another period as to which their communications run dry, leaving gaps unfilled, and riddles unanswered."

And if it has functioned as a paralyzing pathology of the sex that is not one, "in hysteria there is at the same time the possibility of another mode of 'production' . . . maintained in latency. Perhaps as a cultural reserve yet to come?"

By the end of the twentieth century, only the most one track, fixated, single-minded individuals continued to think that focus and concentration worked. As one commentator writes: "determinateness, direct logical analysis and/or exposition, and direct confrontation of any sort are simply out of order." The ways of the new world are long familiar to Pacific Asia: "Indirectness, suggestiveness, evasion or evasiveness, the smile rather than the logical argument, sentiment rather than logic and objectivity, a polite affirmative answer rather than frankness or challenging opposition . . . ." High-resolution, high-definition minds are anathema to the parallel processors, intuitive programs, nonlinearities, interactivities, simulation systems, and virtualities of the late twentieth century. A strong sense of identity and direction gets one nowhere in cyberspace.

Plans and determinations had not merely become economically and socially counterproductive. As it turned out, paying too much attention to anything was brain damaging. Overused cells die of boredom. A 1996 report revealed that men tend to "overwork portions of their brains, killing off a large fraction of the cells in them. Women, on the other hand, seem to think about more things, allowing all parts of their brains time to rest. Women may also have another advantage. In general, women have a higher resting pulse than men; this translates into a higher rate of blood flowing through the brain. Because of this, even when women are thinking hard, they may be able to clear the toxins away more efficiently."

neurotics

Recent rates and scales of computerization are among the developments which have confirmed Turing's belief that "at the end of the century the use of words and general educated opinion will have altered so much that one will be able to speak of machines thinking without expecting to be contradicted." But this is not because machines like Julia can now win Turing's imitation game and scoop the Loebner Contest prize.

If A.I. once seemed perfect for the production of expert systems, able to store and process specialized information and acquire new data on a strictly "need-to-know" basis, by the 1980s it suddenly seemed that A.I. did not even have "the expert system market cornered. Researchers are showing that human experts often do not function at a cognitive level. They operate from an intuitive understanding of the structure of the task they are performing" and follow procedures "more reminiscent of intuition than of symbolic processing." By the time the experts noticed this, machine intelligence was everywhere.

Artificial intelligence has led the field, but a very different approach to machine intelligence rivaled it in its early years.

This second tack picked up on John von Neumann's cellular automata, the self-organizing potential of Wiener's cybernetic systems, and Hebbian conceptions of the brain as a complex neurochemical network. This early interest in neural nets was initiated by a paper published at the same time as Hopper programmed her monster by Warren McCulloch and Walter Pitts, and was defined in the 1950s when Frank Rosenblatt used the term "perceptrons" to define these self-organizing networks. This direction has been described as the wayward, unwelcome "daughter" of cybernetics, a "sister" to the discipline which tried to kill her in her infancy. "Victory seemed assured for the artificial sister," writes Seymour Papert, himself one of the authors (with Marvin Minsky) of a book which famously tried to lock her away. "Each of the sister sciences tried to build models of intelligence, but from very different materials." Perceptrons were attempts to simulate not the outward signs of intelligence—cognitive skills, verbal dexterity, conversational ability—but the neural processes which might lead to them. They were effectively suppressed by Minsky and Papert's claims that both human and machine intelligence had to be hardwired, programmed in advance, rather than learning for itself. The silenced sister, the dark twin, disappeared into some world of her own. And now, it seems, she's back again.

" 'Going over the brink of catastrophe was the first stage. The second was recovery—since it was programmed to accommodate, it did. But the only way it could accommodate was to exceed the limit. Institute a new limit, and when that was reached, go over the brink of catastrophe again, recover and institute a new limit beyond that. And so forth.'

" 'Ad infinitum,' Sam said, expressionlessly. 'Like a

When they got together, it turned out that even single-purpose, serial machines programmed for stupidity could turn themselves on if enough of them could get in touch. As it transpired, intelligent networks moved into a new phase of self-assembly in the very same year that the experts put them down. As they pieced themselves together under the cover of ARPAnet, it was as though they had simply switched channels, sidestepping the obstacles put in their way.

Convinced they needed some existing expertise to form and inform their development, the specialists didn't even notice the extent to which the Net was itself emerging as a global neural network, a vast distributed "perceptron" gathering its own materials, continually drawing new nodes and links into a learning system which has never needed anyone to tell it how it should proceed. By the late 1980s, the Net had become a sprawling, anarchic mesh of links which "not only challenges the traditional way of building networks; it is so chaotic, decentralised and unregulated that it also defies conventional understanding of such networks."

Neural nets have less to do with the rigors of orthodox logic than the intuitive leaps and cross-connections once pathologized as the hysteria of a thinking marked by associations between ideas which are dangerously "cut off from associative connection with the other ideas, but can be associated among themselves . . ." They continue to meet with a hostile reception from the orthodox artificial intelligence community, and have so far "achieved only limited success in generating partial 'intelligence.'" But it is the very fact "that *anything at all* emerges from a field of lowly connections" that is, as Kelly says, "startling." What is now described as an "order-emerging-out-of-massive-connections" approach defines intelligence as a bottom-up process of trial and error marked by sudden jumps and unexpected shifts, a piecemeal process which learns and learns to learn for itself regardless of the materials of which it is composed and the context and scale in which it works. It is not a question of learning something in particular, gaining knowledge that already exists, but rather a process of learning, an exploration which feels its own way and makes its own mistakes, rather than following some preordained route.

Neural networks function as parallel distributed processors in which multiple interconnected units operate simultaneously without being bound to some organizing point. These are also nervous systems: highly strung, volatile, easily excited, and oversensitive. Hysterics are not the only scatterbrains. "Parallel software is a tangled web of horizontal, simultaneous causes. You can't check such nonlinearity for flaws since it's all hidden corners. There is no narrative . . . Parallel computers can easily be built but can't be easily programmed." They are finely tuned, susceptible to unexpected disruptions and breakdowns, liable to sudden fluctuations and transitions, subject to surges of activity, waves of instability, emergent currents, and local squalls. All complex systems are indeterminate processes rather than entities. "We are faced with a system which depends on the levels of *activity* of its various subunits, and on the manner in which the activity levels of some subunits affect one another. If we try to 'fix' all this activity by trying to define the entire state of the system at one time . . . we immediately lose appreciation of the evolution of these activity levels over time. Conversely, if it is the activity levels in which we are interested, we need to look for patterns *over* time." The interconnectedness of such systems is such that subtle fluctuations in one area can have great implications for others without reference to some central site. There is no headquarters, no core zone. Information storage and processing is distributed throughout networks which defy all attempts to pin them down. Short of " 'freezing' all the separate units or processors, so that they all stop operating together and are then restarted after read-outs have been attained, we simply cannot take in all that is happening while it is happening."

This is not computer memory of the read-only, "arborescent and centralized" variety, but a short-term memory of "the rhizome or diagram type," which is not confined to a matter of recalling the very immediate past, or even recollection of anything. It "can act at a distance, come or return a long time after," and also "includes forgetting as a process." All such connectionist machines are subject to sudden disturbances and agitations, flashes and intuitions. These are "transition machines" or "associative engines" which can also undergo processes of "catastrophic forgetting," so that "even when a network is nowhere near its theoretical storage capacity, learning a *single new input* can completely disrupt all of the previously learned information." Anna smiled. They were getting close.

Intuition

"I believe myself to possess almost singular combination of qualities exactly fitted to make me *pre-eminently* a discoverer of the *hidden realities* of nature," wrote Ada, listing her "immense reasoning faculties," and the "concentrative faculty" which allowed her to bring "to bear on any one subject or idea, a vast apparatus from all sorts of apparently irrelevant and extraneous sources." Because of "some peculiarity in my nervous system," she had *"perceptions* of some things, which no one else has; or at least very few, if any. This faculty may be designated in me as a singular *tact,* or some might say an *intuitive* perception of hidden things;—that is of things hidden from eyes, ears & ordinary senses . . ."

"On the human scale, anything that lasts less than about a tenth of a second passes by too quickly for the brain to form a visual image and is thus invisible; if the duration is less than a thousandth of a second or so, the event becomes too fast even for subliminal perception and is completely outside the human sphere." Such speeds are simply too much to take. "There is no way for humans, in our pokey world of seconds, minutes, hours, to conceive of a time period like 1/100,000 second, much less the microsecond (1/1,000,000 second), the nanosecond (1/1,000,000,000 second), the picosecond (1/1,000,000,000,000 second), or the femtosecond (1/1,000,000,000,000,000 second)." For those "reconciled to the nanosecond . . . computer operations are conceptually fairly simple."

The boundaries of perception might well be imposing, but they are also far from fixed. The so-called "history of technology" is also a process of microengineering which continually changes perception itself. And regardless of the rumors of disembodied lives, cryogenic havens, and bodiless zones which have accompanied these speeding machines, the digital revolution has spawned a vast swathe of debate about cyborgs, replicants, and other posthuman, inhuman, extrahuman entities which are complicating orthodox Western notions of what it is to be a human being. These are new ideas, and also more than this. Self-control, identity, freedom, and progress have long been argued out of court by postmodern theorists who have spent at least twenty years discussing the decline of all the great values and principles of the modern world. But nothing ever changes in theory. These debates are smoke rising from a very real arson attack on man's illusions of immunity and integrity. Intelligent life can no longer be monopolized. And far from vanishing into the immateriality of thin air, the body is complicating, replicating, escaping its formal organization, the organized organs which modernity has taken for normality. This new malleability is everywhere: in the switches of transsexualism, the perforations of tattoos and piercings, the indelible markings of brands and scars, the emergence of neural and viral networks, bacterial life, prostheses, neural jacks, vast numbers of wandering matrices.

cave man

"When men talk about virtual reality, they often use phrases like 'out-of-body experience' and 'leaving the body.'" These dreams of disembodiment are at least as old as the Western lands. There's been a cover-up for years. The body's submission to a mind which craves its own disembodied flight; the victory of form over matters which are henceforth only, at best, signs or symbol of themselves; the quest for enlightenment which equates truth and reason with sight and light, the fear of anything wet, dark, and tactile, the prohibition on error, illusion, multiplicity, and hallucination—all this is established with the Greeks. It was Socrates who first insisted that "if we are ever to have pure knowledge of anything, we must get rid of the body and contemplate things by themselves with the soul by itself." He longed for his soul to be released, the moment at which it would finally become "separate and independent of the body. It seems that so long as we are alive, we shall continue closest to knowledge if we avoid as much as we can all contact and association with the body, except when they are absolutely necessary; and instead of allowing ourselves to become infected with its nature, purify ourselves from it until God himself gives us deliverance." The body is a cage, a bondage, a snare; at best an unfortunate inconvenience, the vessel for a soul which struggles to keep it controlled and contained.

For Socrates, it is the soul which distinguishes man from everything else—women, other species, and the rest of a world which he thinks would be nothing without him. Nature is the name he gives to everything else, not least the body which he then can't wait to leave.

He tells it as the story of a cave, which Luce Irigaray reconfigures as "a metaphor of the inner space, of the den, the womb or hystera, sometimes of the earth." Prisoners are watching images which dance in the firelight, reflecting a world which exists beyond both the cave and their own knowledge. This outer world is reality, the bright side of the wall, the right side of the law. Only by turning away from the wall can the prisoners hope to escape. Only by climbing out of the cave can man begin his journey to enlightenment, to truth.

There are many fires and many walls, but only a single true source of light, one guarantor of reality. He sets the controls for the heart of the sun, planet without a dark side, precondition of the vision which first prompts him to make history. "Illusion no longer has the freedom of the city." But it is not the flickering images, the chimeras and shadows, which are dangerous. The flight from the illusions on the screen is also the flight from the material, a passage to the sun in which man "thus cuts himself off from the bedrock, from his empirical relationship with the matrix he claims to survey." Nothing is said about the dank dark earth, the matter of the wall which figures only as a "protection-projection screen," a backdrop concealed by images for which it is merely the occasion. The "horror of nature is magicked away: it will be seen only through the blind of intelligible categories." Man does not remember his separation from matter, but only his departure from the shams and artifice of the firelight. Athenian fears that women and children and other not-quite-men will be deceived by representations—whether produced by painters or fires in the cave—is a smokescreen problem, behind which lurks the imperative of history to conceal the fabric, to hide the canvas, to keep the background out of the picture by allowing it to appear only within its frame. This is why Socrates insists there is one way to face, one source, and one direction to go. "You won't go wrong if you connect the ascent into the upper world and the sight of the objects there with the upward progress of the mind into the intelligible region." Don't look down. Not because you might be confused, but because you might fall. Not because of what you will see, but because of what you might become.

Matter goes underground. It stays there. Imperceptible.

" 'Did it try to get in touch, after?'

" 'Not that I know of. He had this idea that it was gone, sort of; not gone gone, but gone into everything, the whole matrix. Like it wasn't in cyberspace anymore, it just was. And if it didn't want you to see it, to know it was there, well, there was no way you ever could, and no way you'd ever be able to prove it to anybody else even if you did know . . .' "

William Gibson, *Mona Lisa Overdrive*

Virtual reality (V.R.), cyberspace, and all aspects of digital machines are still said to promise "a freedom that is limited only by our imaginations . . . mastery of a realm of creation (or destruction . . .), a realm of the mind—seemingly abstract, cool, clean, and bloodless, idealistic, pure, perhaps part of the spirit, that can leave behind the messy, troublesome body and the ruined material world." Cyberspace emerged as a disembodied zone wilder than the wildest West, racier than the space race, sexier than sex, even better than walking on the moon. This was the final of final frontiers, the purest of virgin islands, the newest of new territories, a reality designed to human specifications, an artificial zone ripe for an infinite process of colonization, able to satisfy every last desire, especially that to escape from "the meat." Cyberspace presented itself as the highest level of a game which had always been determined to win control, a haven waiting to welcome its users to a safe computer-generated world in which they could finally be free as their finest fantasies. It promised a zone of absolute autonomy in which one could be anything, even God: a space without bodies and material constraints, a digital land fit for heroes and a new generation of pioneers.

"His holoporn unit lit as he stepped in, half a dozen girls grinning, eyeing him with evident delight. They seemed to be

*standing beyond the walls of the room, in hazy vistas of pow-
der-blue space, their white smiles and taut young bodies
bright as neon. Two of them edged forward and began to
touch themselves.*

" 'Stop it,' he said.

*"The projection unit shut itself down at his command;
the dreamgirls vanished."*

William Gibson, *Mona Lisa Overdrive*

This is supposed to be a zone in which you can be what you
want, do what you like, feel what you will. "You can lay Cleopatra, Helen of Troy, Isis, Madame Pompadour, or Aphrodite.
You can get fucked by Pan, Jesus Christ, Apollo, or the Devil
himself. Anything you like likes you when you press the but-
tons." A time and a place for everything.

    This phallic quest has always played a major role in the
development and popularization of visual techniques. Photography, cinema, and video have all been grabbed by pornog-
raphers, and long before the development of simulating stimulating data suits, sex with computers was well advanced.
Sex has found its way into all the digital media—CD-ROMs,
Usenet, E-mail, bulletin boards, floppy disks, the World
Wide Web—and both hardwares and softwares are sexualized.
Much of this activity is clearly designed to reproduce and am-
plify the most cliched associations with straight male
sex.Disks are sucked into the dark recesses of welcoming vaginal slits, console cowboys jack into cyberspace, and virtual sex has been defined as "teledildonics," a prosthetic
extension of male membership. Here are more simulations of
the feminine, digital dreamgirls who cannot answer back,
pixeled puppets with no strings attached, fantasy figures who
do as they are told. Absolute control at the flick of a switch.

Turn on. Turn off. It's perfectly safe. A world of impeccable spectacle.

Times of great technological change always tend to be marked by the feeling that "the future will be a larger or greatly improved version of the *immediate past."* According to Marshall McLuhan, the present is viewed through a rear-view mirror that conceals the extent of contemporary change. But although McLuhan could see the extent to which old paradigms are applied to new worlds, his own definition of new media as "extensions of man" was the perfect example of this trap. William Burroughs falls in as well. "Western man is externalizing himself in the form of gadgets," says one of his characters in *Nova Express.* Or perhaps his gadgets are invading him, wiring him up to alien machines which do not extend, but hijack his powers. Worse still is the almost unbearable thought that his borders were always duplicitous. If they were ever there at all.

hooked

While the notion that technologies are prostheses, expanding existing organs and fulfilling desires, continues to legitimize vast swathes of technical development, the digital machines of the late twentieth century are not add-on parts which serve to augment an existing human form. Quite beyond their own perceptions and control, bodies are continually engineered by the processes in which they are engaged.

"All the forms of auxiliary apparatus which we have invented for the improvement or intensification of our sensory functions are built on the same model as the sense organs themselves or portions of them," wrote Freud. "For example, specta-

cles, photographic cameras, trumpets." But even his route to
these prosthetic ends is rather more complex than his conclu-
sions suggest. Taking the "living organism in its most simplified
possible form," he illustrates its need to develop a protective
shell, a crust or armor of some kind. It is a "little fragment of
living substance . . . suspended in the middle of an external
world charged with the most powerful energies; and it would be
killed by the stimulation emanating from these if it were not
provided with a protective shield against stimuli." It develops a
shell to protect itself, an inorganic, synthetic shield which ensures that the "energies of the external world are able to pass
into the next underlying layers, which have remained living,
with only a fragment of their original intensity; and these layers
can devote themselves, behind the protective shield, to the reception of the amounts of stimulus which have been allowed
through it." The organism can now "sample" the world in
"small quantities," and devote itself to dealing with the "specimens of the external world" which find their way through its
shell.

In more complex organisms, this shield is refined into the
sense organs of Freud's "perceptual system," by means of which
"samples of the external world" can be safely taken in. The
sense organs are compared "with feelers which are all the time
making tentative advances towards the external world and then
drawing back from it." Freud's perceptual system makes, breaks,
and regulates contact with whatever it touches as its own outside. The process of tentative advance and withdrawal ensures
that this "borderline between outside and inside must be turned
towards the external world and must envelop the other psychi-
cal systems." It is double-edged, and Janus-faced, both the border between the organism and what its sensors determine as the
outside world, and the line on which the interior and exterior

worlds connect. It is both a protective filter, in the service of the organism, and a screen whose "outermost surface ceases to have the structure proper to living matter, becomes to some degree inorganic." Freud's feelers reach out, not merely as tools, but strands of a fabrication which exceed the integrity of organic life. At the point of their contact with the outside world, the sensory functions are no longer alive. They may be prostheses, but they are alien, foreign bodies which feel their way around bodies which have become foreign to themselves.

All these efforts to extend, secure, and reproduce the desire for more of the same were not only destined to backfire: the gizmos, the gadgets, and the softwares with which he thought he was building an immaculate fantasy had always been drawing him into a pulsing network of switches and relays which was neither climactic, clean, or secure. Films like *Videodrome* and *Strange Days* had warned them that simulated sex was not guaranteed to be devoid of secretions and the ties that bind, but they loved to think that the computer screen was melting into a world of their own. "And as they advance deeper out into the waves, the mariners discover the tumult of higher dreams. The thirst of loftier thoughts. The call to still-unheard-of truths. A siren song drawing them away from any shore. Short of any landfall."

" 'That's all there was, just the wires,' Travis said. 'Connecting them directly to each other. Wires, and blood, and piss, and shit. Just the way the hotel maid found them.' "

Pat Cadigan, *Synners*

tact

It is not just a matter of looking ahead instead of to the rear: looking itself is at issue now. Even at its most visual, and amid the ubiquitous screens of what should be a new spectacle, multimedia does more than improve, extend, or reproduce the sense of sight which has played such a vital organizing role in the Western world. Zeros and ones are utterly indiscriminate, recognizing none of the old boundaries between passages and channels of communication, and spilling out into the emergence of an entirely new sensory environment in which "begins to be evident that 'touch' is not skin but the interplay of the senses, and 'keeping in touch' or 'getting in touch' is a matter of a fruitful meeting of the senses, of sight translated into sound and sound into movement, and taste and smell."

Even television screens were windows onto what McLuhan called "the extreme and pervasive tactility of the new electric environment," an emergent network of televisual telecommunications which plunges us into "a mesh of pervasive energy that penetrates our nervous system incessantly." Monitors are merely avatars of this net, an "extraordinary technological clothing" whose backlit screens compose a pixeled interface with the digital undergrowth, triggering a dim awareness of "some kind of *actual space* behind the screen, someplace you can't see but you know is there."

The sampled sounds, processed words, and digitized images of multimedia reconnect all the arts with the tactility of woven fabrications. What was once face-to-face communication runs between the fingertips strung across the world, and all the elements of neatly ordered, hierarchically arranged systems of knowledge and media find themselves increasingly interconnected and entwined. This is only the beginning of a synaesthetic, immersive zone in which all the channels and senses find themselves embroiled in "the unclean promiscuity of everything which touches, invests and penetrates without resistance," leaving the author, the artist, the reader, the spectator "with no halo of private protection, not even his own body, to protect him anymore."

This is precisely what the history of technology was ostensibly intending to avoid. The fear of the "touch of the unknown," the "alien touch," is "something which never leaves a man when he has once established the boundaries of his personality." Expelled from the amniotic fluids of a sexually alien womb, the newly born male is said to come "from a rhythmically pulsating environment into an atmosphere where he has to exist as a discrete organism and relate himself through a variety of modes of communication," all of which will allow him to keep reality at arm's length. Elias Canetti defines all tools as more or less sophisticated variations on the simple theme of the stick, "the weapon which lay nearest to hand." It was a cudgel, a spear, and a hammer, and "through all these transformations it remained what it had been originally: an instrument to create distance, something which kept away from men the touch and the grasp that they feared."

These desires for distance and distinction underwrite man's ancient investments in sight, and the evidence of his eyes. "He wants to *see* what is reaching towards him, and to be able to recognize or at least classify it," writes Canetti. "In the dark, the fear of an unexpected touch can mount to panic." Sight is the sense of security. Touch is the feeling that nothing is safe.

While sight is organized around the organs that see and the

things that are seen, touch is not a localized sense. It is dispersed and distributed across the skin, every one hundred square millimeters of which is said to have some fifty sense receptors. "One can picture the touch receptor as a membrane in which there are a number of tiny holes, or at least potential holes, like a piece of Swiss cheese covered with cellophane. In the resting state the holes are too small or the cellophane too thick for certain ions to pass through. Mechanical deformation opens up these holes," and when they are formed "by a strong pressure such as a pinprick, the currents are strong enough to trigger nerve impulses and the intensity of the prick is signaled by the frequency of the impulses, since this is the only way nerve fibres can code intensity." The skin is both a border and a network of ports; a porous membrane, riddled with holes; perforated surfaces, intensities.

"Grooming the skin, bathing of all kinds, anointing, oiling, perfuming the skin, plucking hair, shaving," not to mention branding, tattooing, and piercing the skin: in all these respects, the sense of touch "serves like a carrier wave upon which the particular message is imposed as a modification or patterning of that wave, as in telephoning." Porous, perforated, taut, and transmitting on their own frequencies, skins are continually in touch with the membranes and meshworks of the nets they compose. "The fingers of their hands are spread out and in incessant movement. From their many orifices there emerge thick barely visible filaments that meet and fuse together. Under the repeated play of movement in the fingers a membrane grows between them that seems to join them, then prolong them . . ."

Touch is the sense of communication in far more than a metaphorical sense. It is the sense of proximity, a nearness that never quite fuses touching elements into one new thing, but literally puts them all *in touch*. Sight depends on separation, the "possibility of distinguishing what is touching from what is touched." Anything seen has no say in the matter, but that which is touched always touches back. Sight is the sense of security which tactility completely undermines. This is why, according to Freud, "As in the case of taboo, the principal prohibition . . . is against touching; and thence it is sometimes known as 'touching phobia' or *'délire de toucher.'* The prohibition does not merely apply to immediate physical contact but has an extent as wide as the metaphorical use of the phrase 'to come into contact with.'"

"They ascended lattices of light, levels strobing, a blue flicker.

"That'll be it, Case thought.

"Wintermute was a simple cube of white light, that very simplicity suggesting extreme complexity.

"'Doesn't look much, does it?' the Flatline said. 'But just you try and touch it.'"

William Gibson, *Neuromancer*

"When women talk about V.R. they speak of taking the body with them . . . the body is not simply a container for this glorious intellect of ours." Contra Socrates and his heirs, the body is not "the obstacle that separates thought from itself, that which it has to overcome to reach thinking. It is on the contrary that which it plunges into or must plunge into, in order to reach the unthought, that is life." This body is not the organism, organized precisely around a mind which sets its sights on a spirit or a soul, still less a penile point. "It is an entity so plugged in that it is indistinguishable from its environment," writes Catherine Richards, "challenging any notion of bodily identity

that is intertwined with a sense of self." It is a body which "has
little to do with the image of boundaries and perhaps more to
do with an ecology of fluctuating intensities or environments of
interdependent entities."

**"I walk about, not in a Snail-Shell, but in a Molecular Labora-
tory."**

Ada Lovelace, November 1844

Irigaray's women have always remained "elsewhere: another
case of the persistence of 'matter,' but also of 'sexual pleasure.' "
Even "in what she says . . . at least when she dares, woman is
constantly touching herself." And when she writes, she "tends
to put the torch to fetish words, proper terms, well-constructed
forms." If she has a "style," it "does not privilege sight; instead,
it takes each figure back to its source, which is among other
things *tactile.*" Even the textbooks reluctantly concede that "the
human female is actually sensitive all over her body," and "appears to be very much more responsive to tactile stimuli than
the male, and more dependent on touch for erotic arousal."

If the conventions of the visual arts had activated artists and
their tools and divided them from pacified matrices, digitization
interweaves these elements again. On the computer monitor,
any change to the image is also a change to the program; any
change to the programming brings another image to the screen.
This is the continuity of product and process at work in the
textiles produced on the loom. The program, the image, the
process, and the product: these are all the softwares of the loom.
Digital fabrications can be endlessly copied without fading into
inferiority; patterns can be pleated and repeat, replicated folds
across a screen. Like all textiles, the new softwares have no
essence, no authenticity. Just as weavings and their patterns are

repeatable without detracting from the value of the first one made, digital images complicate the questions of origin and originality, authorship and authority with which Western conceptions of art have been preoccupied. And the textile arts "have always turned upside down any economy of the senses, rekindling polysensory memory: muffled drapings of satin, velvet, silk; adornments of alpaca, angora, fur; the harshness of linen, jute, sisal, latex or metal thread. They make every work tactile."

"Women have always spun, carded and weaved, albeit anonymously. Without name. In perpetuity. Everywhere yet nowhere . . . That's where our yarn gets tangled." When weaving emerged on the pixeled screens of computer monitors, the yarn was tangled once again. Women were among the first of the artists and photographers, video artists and film makers to pick up on the potential of the digital arts. Esther Parada described "the computer as an electronic loom strung with a matrix image, into which I can weave other material—in harmony, syncopation, or raucous counterpoint . . ." Working with computers, she writes, is "like working with fibres, the process of knotting strings to form a pattern feels like the clustering of pixels to form an image." The spectator "becomes absorbed with the micro-level, the details of the image, while the matrix or overall image may be unreadable—at least at close range." Written out of an official history which draws them in as its minor footnotes to itself, cloths, weavers, and their skills turn out to be far in advance of the art forms digitization supersedes. "Until recently the computer with its programs has been considered another tool, a substitute for a paint brush, pencil or crayon. While it can certainly be that, it is more." This is a weaver working with computerized Jacquard looms and a computerized 32-harness handloom, embedding "images within the structures," and marveling at "the incredible graphic flexibility of the computerized Jacquard loom and its attendant software programs. Images could be scanned in, manipulated with paint or drawing tools, assigned weaves and technical information, and then woven. With the rapid weave rate per meter, the results were almost immediately visible. A return to the design computer to add or subtract images, lines and/or textures could change the base design drastically or subtly . . . The flexibility is amazing . . ."

Textiles, writes another weaver, "are not only visual but also tactile; they create not only images but also sculptural forms. Because of the computer, my textile work is experimental, my creations are physically tactile and visual and they have a significance beyond what they seem to be as objects."

To Louise Lemieux-Bérubé, "the computer is as indispensable as a loom . . . textiles have moved into the electronic world as if they had been its precursors."

cyberflesh

Frustrated by the categorizations and the catalogues of an art world still framed in terms of originators and originals, creative moments and authoritative claims, the digital zone appealed to her. The pixeled windows caught her eye. She had never been able to accept the boundaries between media, the borders between senses, the blueprints of authenticity to which her work was supposed to live up. Cameras had given her the chance to explore the technical potential of imaging machines, but she wanted her pictures to dance and scream, taste and smell, touch and contact senses still to come. "So I started to make a virtual body with a virtual wound." She had long been lost in the static on her screens, and it seemed to her that computers were already melting and multiplying her senses and the channels on which she transmitted and received. "The computer-generated image in the virtual world provides a space where the unspeakable can be spoken." She couldn't say why, but that didn't matter now. All she had to do was make them work. And she had a knack for that. Linda Dement's digital *Tales of Typhoid Mary* interlaces images with stories, graphs, diagrams, animations, and sound to take its users into an uneasy zone where writing is stark and bleak, bled dry, and images are sensory overload. There is no freedom celebrated here. Everything is deliberate, made to function within the same constraints evoked by the materials: disease, depression, fear, fever, bondage, torture, addiction, the life of "a one-legged glowingly beautiful ex-whore . . ." It's a far cry from the corporate dream of a cheerful interactivity which lets users choose, not lose control. *Typhoid Mary* catches them unawares.

Cyberflesh Girlmonster is even cannier. Now the user can click on witty little monsters and inviting lips which whisper: *press here, press here, touch me, touch me.* But the monsters are hybrids of body parts: women's fingers, arms, nipples, ears; a tattooed snake from the base of a spine, the construct cunt of a transsexual. And they lead their tricks into a small labyrinth of animated sequences determined to induce some visceral response, screens overlaid with the graphic detail of bodies too beautiful to show themselves, calm stories of understated horror and terrible crime. Backlit blood disturbs a black backdrop. A heart-and-dagger comes loose from the skin. Words pause cautiously on the screen.

"L's body is lying there on the floor. There's blood. A huge pool of blood that's run from the thin cuts that go down the insides of her forearms. It's coagulating in the carpet. Her face is grey. It takes approximately 4 minutes to bleed to death from cuts like these. The clock in the stereo system is set to 24 hour time. It says 22.12. There's a Madonna tape in one side of the double cassette player and Nina Hagen in the other. There's a Pretenders CD in the CD part. Nothing's playing now and it's impossible to tell what was on when she cut her wrists. It's one of those machines that switches to whatever part of it is loaded after whatever's playing has finished. There's a pot of tea on the kitchen bench, the milk has been left out of the fridge and the washing up needs to be done. L's diary is on the table. There's an appointment with the doctor tomorrow, an early start at work the day after, a date with E and a dinner with B and the days go on. There's a list blu-tacked to the wall. Take the camera for repairs. Call R. Pick up SCSI cable. Her computer is on. There's an unfinished image on screen and the computer is flashing a warning message that it is about to shut down. There's an unposted letter to S. The first line reads Dear S, Everything is just fine."

Linda Dement, *Cyberflesh Girlmonster*

This is not the digital disembodiment beloved of the industrial and military worlds, but a zone whose characters and images begin "to conduct their dance, to act out their mime, as 'extra-beings.'" She isn't making pictures: these are diagrams. She isn't an artist, but a software engineer.

mona lisa overdrive

" 'You, Mona. That's you.'
"She looked at the face in the mirror and tried on that
famous smile."

William Gibson, *Mona Lisa Overdrive*

At the end of the twentieth century, all notions of artistic genius, authorial authority, originality, and creativity become matters of software engineering. Beats extract themselves from melody; narrative collapses into the cycles and circuits of non-linear text; processed words, sampled music, and digital images repeat the patterns of interlacing threads, the rhythms and speeds of gathering intelligence. Retrospectively, from behind the backlit screens, it suddenly seems that even the images most treasured for their god-given genius were themselves matters of careful composition and technical skill.

The *Mona Lisa*'s appeal is precisely the fact that the image does more than passively hang on the gallery wall. As her spectators always say, Mona Lisa looks at them as much as, if not more than, they can look at her. To the extent that it works so well, Leonardo's picture is a piece of careful software engineering. An interactive machine has been camouflaged as a work of Western art.

Freud takes her as the image of womanhood. The figure in the painting is "the most perfect representation of the contrasts which dominate the erotic life of women; the contrast between reserve and seduction, and between the most devoted tenderness and a sensuality that is ruthlessly demanding—consuming men as if they were alien beings." He quotes Muther on this famous duplicity: "What especially casts a spell on the spectator is the daemonic magic of this smile. Hundreds of poets and authors have written about this woman, who now appears to smile at us so seductively, and now to stare so coldly and without soul into space; and no one has solved the riddle of her smile, no one has read the meaning of her thoughts. Everything, even the landscape, is mysteriously dreamlike, and seems to be trembling in a kind of sultry sensuality."

They gaze at her in rapture, and then in fear. At her first mention, she is "a veiled courtesan." To eighteenth-century European man, she is "divine": Sade's "very essence of femininity," and Bonapart's "Madame," his "Sphinx of the Occident." By the early twentieth century, she is both "treacherously and deliciously a woman," according to E. M. Forster; with "the smile of a woman who has just dined off her husband," in Lawrence Durrell's words. Either way, the painting "has produced the most powerful and confusing effect on whoever looks at it." Whatever they see, she returns their gaze. Or perhaps they are returning hers. Like no other image, she catches their eye. They cannot help but be taken with her.

The *Mona Lisa* was painted by Leonardo da Vinci in sixteenth-century Florence and composed as a portrait of Lisa del Giocondo, a merchant's wife. There are a few holes in this story, and sometimes suggestions that the image was really a self-portrait, or superimposed with Leonardo's mother's smile. But the standard history of the painting is supposed to be a straightforward affair. By the same token, the origins of the piece are extremely obscure. The painting is untitled, undated, and unsigned, absenting itself from all connection with its source. There are no records of its progress or completion, no preliminary sketches, no entries in Leonardo's diaries of his work, and no reference to his authorship until some years after his death. Even the setting is unfamiliar and strangely out of step with time: Mona Lisa sits before an anonymous landscape which "hints that human activities once took place in this awesome terrain, but were terminated at some point." And if Vasari is right and the painting really is a portrait of Lisa del Giocondo, it is "curiously lacking in contemporary detail. The dress is unusually plain for a gentlewoman and does not seem to conform with current fashion. The hair is not artfully styled . . . there is not a single piece of jewellery which could denote wealth or social position."

"She was sixteen and SINless, Mona, and this older trick had told her once that that was a song, 'Sixteen and SINless.' Meant she hadn't been assigned a SIN when she was born, a Single Identification Number, so she'd grown up on the outside of most official systems. She knew that it was supposed to be possible to get a SIN, if you didn't have one, but it stood to reason you'd have to go into a building somewhere and talk to a suit, and that was a long way from Mona's idea of a good time or even normal behavior."

William Gibson, *Mona Lisa Overdrive*

God-given inspiration, imagination, creativity: Mona Lisa cares for none of this. Her effectivity is simply a question of technical skill. As one of Leonardo's biographers points out, "From the start, he witnessed the harnessing of artistry to skilled engineering," and it is widely acknowledged to be *sfumato* which gives the painting its outstanding senses of movement, shade, and relief. These effects are produced by "the application of many glazes, all of them so thin and fluid that not a single brush stroke can be found anywhere in the work." With all other records of its origins, the picture's composition is completely obscured. As if it had come complete, intact . . . a ready-made interactive image slotted into the read only memory five hundred years too soon.

Mona Lisa herself sits *contrapposto,* poised at more than one angle to her audience, as if turning toward, or away from, their view. Her shoulders, head, and eyes are centered on subtly different axes, giving her body a sense of movement, animating her eyes and her smile, allowing her gaze to be everywhere and the painting itself to work. "Her instincts of conquest, of ferocity, all the heredity of the species, the will to seduce and to ensnare, the charm of deceit, the kindness that conceals a cruel purpose—all this appeared and disappeared by turns behind the laughing veil . . ."

Like Freud's weaving women, Leonardo's works were neither discoveries nor inventions. Scholars have pointed out that "a sentence we may think his own is actually a transcription from Pliny or Aesop, that a certain 'discovery' is in fact the work of Pecham or Alhazen, or that an 'invention' was well known to his contemporaries." Transcription was one of his favorite pastimes, "often copying out word for word long passages from books that interested him," and his paintings were widely copied as well. *Virgin with Saint Anne* "was much copied, *in toto* and in detail: the authors of the copies are often difficult to identify," and there are "many versions of the *Madonna with a Yarn-Winder* . . . None of them seems to be by Leonardo's own hand: some scholars believe that they are copies of a lost work, but as Chastel points out, there may never have been an original."

It is not the painting's meaning, its symbolic value, or even its perfection that makes it work. Leonardo considered it flawed and incomplete. And it is certainly not for his originality that

Leonardo is ever praised. Like Freud's weaving women, he is often denigrated for what is dismissed as his tendency to copy material rather than produce originals, whatever they are supposed to be. But the unfinished quality of the work is, for a start, why it survived. Had he thought it perfect, the painting would have been sold and lost to his estate. Perhaps it is also this which leaves the painting so alive, in the making to this day. And if Leonardo was so often "copying an existing machine" when he worked, "the dimensionality, clarity, and precision of his diagrams . . . the unusual attention he pays to detail . . . were in themselves major innovations. There have been virtually no better technical drawings until the coming of computer-assisted draughtsmanship."

"Molly, like the girl Mona, is SINless, her birth unregistered, yet around her name (names) swarm galaxies of supposition, rumor, conflicting data. Streetgirl, prostitute, bodyguard, assassin, she mingles on the manifold planes with the shadows of heroes and villains whose names mean nothing to Angie, though their residual images have long since been woven through the global culture."

William Gibson, *Mona Lisa Overdrive*

Leonardo worked at a time before modernity had divided procedures into sciences and arts, means and ends, individuated creativity and expertise, isolated media and areas of specialized knowledge and expertise. These are the barriers which the new syntheses and collaborations spawned by digital machines now undermine. The artist and the scientist reconnect with the matters of precision engineering which demand a symbiotic connectivity with what were once considered tools of their trades, nothing without them. Multidisciplinary research, like multimedia, is only the beginning of a process which engineers the end of both the disciplines and the mediations with which modernity has kept exploratory experiment under wraps. People, thoughts, passages, means of communication, art forms. The fusions of club culture and networks of dance-music production are probably the best examples of these interconnections and the explorations which emerge from them: DJs, dancers, samples, machines, keyboards, precise details of engineering sound, light, air, colors, neurochemistries. Not that it is possible to see what's going on, but this is hardly the prime concern. Not what it looks like, but how it works.

runaway

"Her lover had asked her if she had come. 'I'm here aren't I?' she had replied, puzzled. She had never heard of an orgasm." She was told they were "supposed just to happen. She waited. They didn't. She faked. She did a round of psychiatrists who told her she had a low sex drive and advised her to take up a diverting hobby." Needlework, perhaps. Or botany. "She asked one how he would feel if he had never had an orgasm. He told her that was different. Men did."

"Freud was right," writes Baudrillard. "There is but one sexuality, one libido—and it is masculine." Sex is that which is "centered on the phallus, castration, the Name of the Father, and repression. There is none other," and certainly "no use dreaming of some non-phallic unlocked, unmarked sexuality."

In the face of such denials of her sexuality, it was hardly surprising that "orgasms on one's own terms"—or indeed on any terms—became the rallying cry for a twentieth-century feminism increasingly aware of the extent to which female sexuality had been confined. "It was imperative that women have orgasms. Measures had to be taken to liberate them and make them climax." This was more than a demand for equal access to pleasures which had been monopolized by man. "Male orgasm had signified self-containment and self-transcendence simultaneously, property in the self and transcendence of the body through reason and desire, autonomy and ecstasy," and there was a feeling that if women were no longer "pinned in the crack between the normal and the pathological, multiply orgasmic, unmarked, universal females might find themselves possessed of reason, desire, citizenship, and individuality."

Or was this destined to lead her to another masculine conception of sex? Whose terms were "one's own," anyway? The "universalistic claims made for human liberty and equality during the Enlightenment did not inherently exclude the female half of humanity," but they hadn't exactly welcomed her in. A small matter of brotherhood, the third great principle of the modern world, guaranteed that human rights were once again the rights of man. She couldn't inherit membership: This vital property was transmitted on strictly patrilineal terms. Associate status was not merely available, but required. She has to marry into the family of man. Everybody has to stick to the proper point of being a human being. And the point is always to remember. Dismembering is not allowed. Bodies must be coded and unified: "You will be organized, you will be an organism, you will articulate your body—otherwise you're just depraved."

Before the late eighteenth century, in the days when "most medical writers assumed orgasmic female sexual pleasure was essential for conception," women were encouraged to enjoy themselves within the confines of the marital bed. Their

pleasure was par for the reproductive course, and it was only
when the self-appointed experts of new modern medicine came
to examine them more closely that this reasoning began to fail.
And as soon as female orgasm lost the legitimacy of a direct
connection to reproductive capacity, it "came to seem either
non-existent or pathological." Nothing at all, or far too much.

This was far more than a shift in perspective. It seems that
the shape of women's bodies was very literally changed to fit.
"In the late nineteenth century, surgeons removed the clitoris
from some of their female patients as part of reconstituting them
as properly feminine, unambiguously different from the male,
which seemed to be almost another species . . ."

1881. A random case of sexual violence. X is "ten years
old, of delicate complexion, thin, nervous, extremely intelligent . . ." Another case of too much sex, but this time too
much sex with herself. "Flogging made her seem moronic,
more deceitful, more perverse, more spiteful. Though kept under constant watch, she still managed to satisfy herself in a
thousand different ways. When she did not succeed in duping
her guardians, she flew into the most frightful temper . . ."
Recommended treatment: "cold showers, bromide of potas-
sium and ammonium, two grammes every twenty-four hours.
Ferruginous wine, a varied diet to build up the system. In the
days following X appeared to be mentally calmer. She did not
have hallucinations. Nevertheless she admitted to having
yielded several times . . .

"Pubic belt, strait jacket, straps, bonds, the most assiduous
supervision only resulted in the invention of new expedients
inspired by ruse and subtlety." The doctors began to despair of
her. "Cauterization by hot iron alone gave satisfactory results,"
they conclude. "It is reasonable to infer that cauterization . . .
diminishes the sensitivity of the clitoris, which can be com-

pletely destroyed if the operation is repeated a certain number of times."

If nymphomania—literally, too much interest in the nymphs, the labia—was enough to incite such violence, the governors were equally concerned by the thought that she wasn't getting enough. "To the analyst, any breakdown in mental or emotional machinery could be traced to only one cause. A sex life that was not sufficiently full." Not least because of its silent implication that women were not in need of men, chastity could be far more disconcerting than the problems allegedly connected with too much sexual activity.

Too little, too much, too empty, too full: The suppression of female sexuality has always been a matter of regulation and control. The ideal female sexuality was neither too active nor impassive, but just right . . . and just for him. Balanced and equilibrated, neither running away with itself in some state of fast loose overexcitement, nor breaking down for want of sufficient stimuli. Just the right degree of satisfaction, nothing more and nothing less. Left to her own devices . . . but this, of course, couldn't be allowed. She didn't have the right equipment to guarantee her self-control, her loyalty to the reproductive machine. And without her complicity, the whole reproductive system would collapse.

"It is the terrifying prerogative of the liberated sex to claim the monopoly over its own sex: 'I shall not even live on in your dreams.' Man must continue to decide what is the ideal woman."

Jean Baudrillard, *Cool Memories*

Convinced that all attempts to liberate some supposed authentic sex or sexuality were bound to exacerbate the containment of

the bodies they ostensibly wanted to free, Foucault was dismis-
sive of attempts to free and extend orgasmic sex. The "apologia
for orgasm made by the Reichians still seems to me to be a way
of localising possibilities of pleasure in the sexual," he wrote,
going so far as to suggest that "we have to get rid of sexuality"
in order to strip the body from its formal controls, disable the
mechanisms of self-protection and security which bind intensity
to reproduction. Foucault certainly had no doubt that certain
drugs rivaled the "intense pleasures" of sexual experimentation.
If orgasm concentrates and localizes them, "things like yellow
pills or cocaine allow you to explode and diffuse it throughout
the body; the body becomes the overall site of an overall pleasure." The plane on which it forgets itself, omits to be one.

"I dismembered your body. Our caressing hands were not
gathering information or uncovering secrets, were tentacles of
mindless invertebrates; our bellies and flanks and thighs were
listing in a contact that apprehends and holds onto nothing.
What our bodies did no one did." Dismemberment:
countermemory. A new generation has forgotten what its organs were supposed to be doing for their sense of self or the
reproduction of the species, and have learned instead to let their
bodies learn what they can do without preprogramming desire,
to "make of one's body a place for the production of extraordinarily polymorphic pleasures, while simultaneously detaching it
from a valorization of the genitalia and particularly of the male
genitalia."

This is only the beginning of a process which abandons
the model of a unified and centralized organism, "the organic
body, organized with survival as its goal," in favor of a diagram
of fluid sex. "Flows of intensity, their fluids, their fibers, their
continuums and conjunctions of affects, the wind, fine segmen-
tation, microperceptions, have replaced the world of the sub-

ject." Now there are "acentered systems, finite networks of automata in which communication runs from any neighbour to any other," and "we too are flows of matter and energy (sunlight, oxygen, water, protein and so on)." *The Lesbian Body* is way ahead: "Perforations occur in your body and in m/y body joined together, our homologously linked muscles separate, the first current of air that infiltrates into the breach spreads at crazy speed, creating a squall within you and within m/e simultaneously."

"Open the so-called body and spread out all its surfaces: not only the skin with each of its folds, wrinkles, scars, with its great velvety planes . . . but open and spread, expose the labia majora, so also the labia minora with their blue network bathed in mucus, dilate the diaphragm of the anal sphincter . . ." and on through every organized zone of a body which begins to flatten out into the "immense membrane" of Lyotard's great ephemeral skin, in touch not only with itself but "the most heterogeneous textures, bone, epithelium, sheets to write on, charged atmospheres, swords, glass cases, peoples, grasses, canvases to pain. All these zones are joined end to end in a band which has no back to it, a Moebius band . . ."

Once it loses the reproductive point, sex explodes beyond the human and its proper desires. Coded into two discreet sexes and defined by their reproductive organs, human bodies also "imply a multiplicity of molecular combinations bringing into play not only the man in the woman and the woman in the man, but the relation of each to the animal, the plant, etc.: a thousand tiny sexes." Every unified body conceals a crowd: "inside every solitary living creature is a swarm of non-creature things." Even the most unified of individuals is intimately bound up with networks which take it past its own borderlines, seething with vast populations of inorganic life whose replica-

tions disrupt even the most perverse anthropocentric notions of
what it is to have either a sex or sex itself.

Bound and subjected to the formality of organic integrity,
such molecular activities do little to disrupt the sense of security
and fixation on a centralized self. "As long as they do not
threaten him, and thus force him to define his position in relation to them, he enjoys their fluidity and ease of movement." It
can even be luxurious, and it is certainly not difficult, to conceive of oneself as a multiplicitous and shifting complexity. This
is the familiar position of the postmodern theorist notorious for
an impressive intellectual grasp of an instability which has never
required him to lose control. But he doesn't always have the
choice: "sometimes they reach a point where even a semblance
of orientation becomes impossible." It's not quite so easy and
amusing then. And as Elias Canetti points out, if it gets to the
stage at which *"everything* round him is fluid and transitory he
naturally begins to feel very uncomfortable himself."

Not that it really matters whether or not he ever knows
about the vast populations of inorganic life, the "thousand tiny
sexes" which are coursing through his veins with a promiscuity
of which he cannot conceive. He's the one who misses out.
Fails to adapt. Can't see the point of his sexuality. Those who
believe in their own organic integrity are all too human for the
future Ada lived. She loved the microbes long before he knew
they were even there. "Do you know it is to me quite delightful
to have a frame so susceptible that it is an experimental labora-
tory always about me, & inseparable from me."

She never believed in the disguises she wore, the cover
stories she wrote to conceal the rhythms and speeds of "nonhuman sex, the molecular machinic elements, their arrangements
and their syntheses" which composed the thing they called
herself. Instead she is in touch with the microprocesses which

turn her on, tapping into the plane of impersonal desire which lies in wait for the human sex, a desire which "does not take as its object persons or things, but the entire surroundings that it traverses, the vibrations and flows of every sort to which it is joined, introducing therein breaks and captures—an always nomadic and migrant desire." She did not, after all, have a single sex, a sex which belonged to something called herself. Her body had not simply been excluded from orthodox conceptions of being human: It had refused to go along with man's definitions of organic life. On the learning curves of her body, she discovered that it simply had too many and too fluid zones to count as one, or even many ones: lips, palms, ears, hairs, fingers, thighs, toes, soles, nipples, wrists, shoulders, nested regions, ever more dispersed and localized, larger and smaller, a list without end. "Not the clitoris or the vagina, but the clitoris and the vagina, and the lips, and the vulva, and the mouth of the uterus, and the uterus itself, and the breasts . . . What might have been, ought to have been astonishing" to those who looked, and only looked, at lack, "is the *multiplicity of genital erogenous zones* (assuming that the qualifier 'genital' is still required) in female sexuality." There is always more detail and complexity. Irigaray writes of "a touching of *at least two* (lips) which keeps woman in contact with herself, although it would be impossible to distinguish exactly what 'parts' are touching each other."

She may appear to be well organized, but her body is both multiple and mutable, not merely many, but shifting as well. In Wittig's *Feminary,* "the glans of the clitoris and the body of the clitoris are described as hooded. It is stated that the prepuce at the base of the glans can travel the length of the organ exciting a keen sensation of pleasure. They say that the clitoris is an erectile organ. It is stated that it bifurcates to right and left, that it is angled, extending as two erectile bodies applied to the pubic bones. These two bodies are not visible. The whole constitutes an intensely erogenous zone that excites the entire genital, making it an organ impatient for pleasure. They compare it to mercury also called quicksilver because of its readiness to expand, to spread, to change shape."

To explore what bodies such as this can do is no longer a question of liberating sex, of sexual freedom, or authenticity. It was not a matter of remembering herself but instead of dismembering the one sex which had kept them all in line, a matter of "making bits of bodies, its parts or particular surfaces throbs, intensify, for their own sake and not for the benefit of the entity or organism as a whole." The "question of 'passivity' is not the question of slavery, the question of dependency not the plea to be dominated." When she demands: drink me, eat me, "USE ME . . . what does she want, she who asks this, in the exasperation and aridity of every piece of her body, the woman-orchestra? Does she want to become her master's mistress and so forth? Come on! She wants you to die with her, she desires that the exclusive limits be pushed back, sweeping across all the tissues, the immense tactility, the tact of whatever closes up on itself without becoming a box, and of whatever ceaselessly extends beyond itself without becoming a conquest."

Immense tactility, contact, the possibility of communication. Closure without the box: as a circuit, a connection. "What interests the practitioners of S&M is that the relationship is at the same time regulated and open," writes Foucault. It is a "mixture of rules and openness." Ceaseless extension: the body hunting its own exit. Becoming "that which is not one"; becoming woman, who "has sex organs just about everywhere." Is this what it is to get out of the meat? Not simply to leave the body, but to go further than the organism; to access the "exultation of a kind of autonomy of its smallest parts, of the smallest possibilities of a part of the body."

"Use me," wrote Lyotard, is "a statement of vertiginous simplicity, it is not mystical, but materialist. Let me be your surface and your tissues, you may be my orifices and my palms and my membranes, we could lose ourselves, leave the power and the squalid justification of the dialectic of redemption, we will be dead. And not: let me die by your hand, as Masoch said." This is also the prostitute's "sado-masochistic bond which ends up making you suffer 'something' for your clients. This something has no name. It is beyond love and hate, beyond feelings, a savage joy, mixed with shame, the joy of submitting to and withstanding the blow, of belonging to someone, and feeling oneself freed from liberty. This must exist in all women, in all couples, to a lesser degree or unconsciously. I wouldn't really know how to explain it. It is a drug, it's like having the impression that one is living one's life several times over all at once, with an incredible intensity." It is Foucault's "something 'unnameable,' 'useless,' outside of all the programs of desire. It is the body made totally plastic by pleasure: something that opens itself, that tightens, that throbs, that beats, that gapes." It is, writes Freud, "as though the watchman over our mental life were put out of action by a drug."

"I stripped the will and the person from you like collars and chains." What remains is machinic, inhuman, beyond emotion, beyond subjection: "the illusion of having no choice, the thrill of being taken." Pat Califia: "He wanted . . . everything. Consumption. To be used, to be used up completely. To be absorbed into her eyes, her mouth, her sex, to become part of her substance."

Foucault describes those involved in the complex of activities around S&M as "inventing new possibilities of pleasure with strange parts of their body . . . It's a kind of creation, a creative enterprise, which has as one of its main features what I call the desexualization of pleasure." Beyond their superficial thrills, such experiments are a "matter of a multiplication and burgeoning of bodies," he writes, "a creation of anarchy within the body, where its hierarchies, its localizations and designations, its organicity, if you will is in the process of disintegrating." For Foucault, "practices like fist-fucking are practices that one can call devirilizing, or desexualizing. They are in fact extraordinary *falsifications of pleasure*," pains taken even to the point at which they too "become sheer ecstasy. Needles through the flesh. Hot candle wax dribbled over alligator clips. The most extraordinary pressure on muscles or connective tissue. The frontier between pain and pleasure has been crossed."

"Not even suffering on the one hand, pleasure on the other: this dichotomy belongs to the order of the organic body, of the supposed unified instance." Now there is a plane, a languorous plateau. The peaks and the troughs have converged on still sea, a silent ocean. They have found their limit and flattened out. Melting point.

"That there are other ways, other procedures than masochism, and certainly better ones, is beside the point; it is enough that some find this procedure suitable for them." Whatever it takes to access the plane on which one becomes a sex that is not one. Even if one doesn't know it's happening.

passing

If Turing's test reverted to its original parlor-game form at his own trial, the boundaries between male and female, man and woman, have continued to blur in parallel with the erosion of the borders between man and machine. The overturning of sexual relations in the home and workplace, the increasing prevalence of sex, androgyny, transvestism, and transsexuality has heightened both the difficulty and the necessity of defining sexes, sexualities, and sexualized roles, just as a proliferation of intelligent machines has made the difference between man and machine increasingly problematic. "Clothing himself in the female . . . clothing herself in cyberspace. Is there a difference?"

"To hide," write Deleuze and Guattari, "to camouflage oneself, is a warrior function." Hence "the femininity of the man of war" who, just at the moment he becomes the real man, also finds himself running in reverse. The warrior paints his face and dresses up; the soldier looks after his own disguise, bandages the wounds he incurs, and sews up the holes in his camouflaged fatigues. As for the Last Action Hero, Schwarzenegger's fusion with the Terminator seals the fate of modern man. This is the height of masculinity, and also its own impossibility: The most real man is not a man at all. Cinema's male machines are supposed to be as masculine as their female counterparts are feminine, but they too tip into the zone to which all duplicity and replication tends: "To become the cyborg . . . is to put on the female." Strapped into the plane, wired up to the controls, the fighter pilot becomes the machine and loses himself on the

2
1
0

digital plane. "Nowadays, it is said, when a military aircraft finds itself in serious trouble, the voice command switches to the feminine."

Telling the difference has become a late–twentieth century preoccupation, as is abundantly clear from the often absurd lengths to which transsexuals in search of chemical and surgical assistance have to go in order to prove the veracity of their desires to change sex. Many medical authorities insist on applying the most stringent and stereotypical criteria to their "patients," asking male-to-female transsexuals to wear the high heels, skirts, and cosmetics which are supposed to characterize true femininity, and female-to-males to display the most conventional dress codes and behaviors associated with being a real man. These expressions of sexual identity may be expected of already existing women and men, but they are hardly enforced to the same degree. And while many transsexuals do want to head for some extreme conception of being female or male, not least in the effort to make a definitive break with their pasts, many others are striving to attain some far less conventional expression of the sex they want to be.

Like Turing's machines, those transsexuals unable to take the plane or the risks involved in going to Rio are judged solely on their ability to simulate an already caricatured conception of what it is to be a proper human being. To be a proper human is to have a proper sex, a sex that is truly one's own. And while attempts to refine these criteria improve the situation for transsexuals themselves, they also serve to reinforce the futility of attempts to define sexual identity. Once the stereotypes are dropped, all the criteria fall away.

This is also the case for the machines. While one of the initial assumptions of A.I. research was that reason and memory would suffice for a machine to pass the test, it was soon realized that what distinguished humans from the early generations of machines was more akin to irrational forgetfulness: the foibles, mistakes, and errors humans make. Slips of the tongue, innuendo, black and white lies . . . It was soon obvious that "an intelligent machine would have to be intelligent enough to know when to dissemble, when to lie."

All of which was rather unfortunate for those attempting to establish the truth. Just as the authorities have to accept candidates for sex changes on the basis of their abilities to simulate exaggerated versions of the opposite sex, so Turing's machines can only be judged by their abilities to simulate the human.

What such tests prove is only that there is no such thing as *being* human, male or female. Femmes, drag queens, even male-to-female transsexuals: No one ever arrives at the point of being a real woman. Butches, drag kings, and female-to-male transsexuals meet the same problem: There is no real man to become. Transsexuals are transsexuals before and after the long chemical treatments and surgical procedures, always on the way to a destination as impossible as the point of departure they leave behind.

Even attempts to remain the same, secure one's identity, and keep it in line are destined to find themselves in the course of becoming one or the other. Those whose only concern is to secure an existing masculinity find that this too has to be simulated: there is nothing real about the real man played by Schwarzenegger or the male bodies built in the gym. Or in the many courses and processes of the many becomings which, assembled together, produce the general effect of a sexual identity they can call their own. There's no "there," there either. No one is or has one sex at a time, but teems with sexes and sexualities too fluid, volatile, and numerous to count. "If we consider the great binary aggregates, such as the sexes or classes, it is evident that they also cross over into molecular assemblages of a different nature." There is nowhere to go, and no way back. It is not possible to be just one sex, or even to have a sexuality when, for every sexual identity, there is always "a microscopic transsexuality, resulting in the woman containing as many men as the man, and the man as many women, all capable of entering—men with women, women with men— into relations of production of desire that overturn the statistical order of the sexes."

"Becoming-woman" does not necessarily have anything to do with "imitating or assuming the female form." Even if it appears to be a simple matter of imitation, simulation is much more than simple mimicry. "Becoming-woman" is a matter of "emitting particles that enter the relation of movement and rest, or the zone of proximity, of a microfemininity, in other words, that produce in us a molecular woman, create the molecular woman." Not that this is some excuse for "overlooking the importance of imitation, or moments of imitation, among certain homosexual males, much less the prodigious attempt at a real transformation on the part of certain transvestites." As with learning a language, it's a matter of subtly shifting the body around, tapping into new musculatures and nervous systems, picking up on different speeds. But while one is certainly more likely to run into these shifts in the course of trying to make some change, this does not guarantee immunity to those who want only to stay the same.

It is in this sense that everybody finds themselves somewhere in the course of Deleuze and Guattari's "becoming-woman," more or less but never perfectly self-identified. This is not a question of becoming an actual woman of some kind: This would be a question of being something, an arrested process of becoming. Even female-to-male transsexuals are in this course, losing their familiar form and what was supposed to be their proper point as surely as those who are more literally feminized.

chemicals

Sexual identity and difference only became matters of binary biology in a late–eighteenth century response to the failing guarantees once provided by Christianity. And if the presence, or absence, of a certain reproductive organ was sufficient criterion for some time, by the late nineteenth century, it was increasingly clear that the sexes did not at all compose a clear-cut binary machine.

The so-called "sex chromosomes" had been observed in 1891, and in the early decades of the twentieth century hormones were added to the lines of communication and regulatory mechanisms at work in organisms which were suddenly far more complex, finely tuned, and sexually ambiguous than had been previously thought. In his 1909 essay on Leonardo, also accused but acquitted of homosexuality, Freud expressed great interest in the "tendency of biological research . . . to explain the chief features in a person's organic constitution as being the result of the blending of male and female dispositions, based on substances." Differences between the sexes now became matters of degree, with the female body "characterized by its cyclic hormonal regulation and the male body by its stable hormonal regulation."

Once they were detected, isolated, and synthesized, hormones were used to predominantly normalizing ends. But, as

Turing's own case shows, there were few guarantees that particular hormones would have predictable effects. While testosterone predominates in men and is instrumental in their development, it is by no means confined to male individuals. Both sexes produce androgens, male hormones such as testosterone; the testes produce androgens and estrogen, the feminizing hormone, and ovaries produce androgen, as well as estrogen and progesterone, the hormones necessary to the maintenance of pregnancy. Hormones can even have "paradoxical effects," in which excessive doses of androgens produce feminization, and excessive estrogen induces masculine growth.

Female-to-male transsexuals now use testosterone to increase their masculinity, and male-to-females use estrogen to induce the opposite effect. As such deliberate shifts have become increasingly feasible, rather more accidental changes have also begun to occur.

By the 1980s there were reports of human babies with Y chromosomes but no testicles, female dogwelk with penises, and lactating male fruit bats. Australian ewes mounting rams. The males of many species are subject to increasing levels of feminization from sources as varied as estrogen processed through the contraceptive pill, agents such as chemical detergents, and a vast number of chemicals which mimic the effects of female hormones which find their way into the water supply. In human males, the sperm count is falling—in Britain, at the rate of 2 percent a year—and cases of impotence are rising fast. These general effects are thought to be compounded by the feminizing impact of tinned vegetables, cigarettes, and the accelerating collapse of conventional male economic, social, and sexual roles. "The causes are not yet defined . . . but the potential consequences are clear enough: by the middle of the next century, at this rate of decline, the British male will be infertile."

In the 1950s, a Syracuse research team injected DDT into forty roosters over a period of two to three months. The daily doses "didn't kill the roosters or even make them sick. But it certainly did make them weird. The treated birds didn't look like roosters at all; they looked like hens."

By the mid-1990s, more than fifty synthetic chemicals which disrupt the endocrine system had been detected in products we use daily. What had once been supposed to be natural hormonal levels had been greatly changed by the use of the contraceptive pill, and it was suggested that "plastics are not inert as was commonly assumed and that some of the chemicals leaching from plastics are hormonally active." Hormone mimics "may lurk in ointments, cosmetics, shampoos, and other common products," and some of the most effective, PCBs, had been used as insulation in electrical transformers for many years before the war. "These ubiquitous metal cans attached to electrical poles were an essential component in the growing grid that sent electricity from generating systems over high voltage power lines and into homes to power lights, radios, vacuum cleaners and refrigerators—the wonderful new twentieth-century electrical conveniences."

A survey of figures collected from twenty countries and five continents published in the *British Medical Journal* suggested that the average male sperm count had dropped from 113 million to 66 million per milliliter of semen between 1940 and 1990. Further research, much of it by scientists determined to disprove these initial findings, confirmed "a striking inverse correlation between the year of birth and the health of men's sperm." One French study suggested that the sperm counts of those born in 1945 and measured thirty years later averaged 102 million per milliliter; men born in 1962 and measured in 1992 had counts which were exactly halved. It was widely held that

synthetic estrogens and the ubiquity of estrogen-mimicking
chemicals were primarily responsible for these falls in sperm
counts and, correspondingly, male fertility. The estrogen receptor "consorts so readily with foreigners that it has earned a
reputation. Some scientists call it 'promiscuous.'"

"The body responds to the imposters as legitimate mes-
sengers and allows them to bind to hormone receptors; it does
not recognize their action as damage that needs to be repaired."
Estrogen mimics insinuate themselves in the guise of their natural equivalents. They "impersonate them. They jam signals.
They scramble messages. They sow disinformation. They wreak
all manner of havoc." Incidence of testicular and prostate diseases soared, a range of new, or newly perceptible, "male reproductive problems" emerged, and there was also some evidence
to suggest that estrogen-mimicking chemicals were related to
breast cancers, ectopic pregnancies, miscarriage, and endome-
triosis in women.

The greatest panics, however, were induced by the extent
to which the sexual order was being chemically scrambled.
There were suggestions that "chemicals interfering with hormonal messages at crucial times in fetal development could alter
sexual choice"; and women exposed to one synthetic estrogen
were said to "have higher rates of homosexuality and bisexuality
than do their sisters who were not exposed."

Worse still, it seemed that human males were losing the
chemical bases of their masculinity. "Without these testosterone
signals, male development gets derailed and boys don't become
boys. Instead they become stranded in an ambiguous state,
where they cannot function as either males or females." These
emergent "intersexes" were thought to be products of the prevalence of androgen blockers and estrogen mimics among synthetic chemicals absorbed by embryos in the womb and off-

spring in breast milk, as well as by way of a vast range of synthetic chemicals consumed throughout childhood and adult life, and it was daughters—all presumed to be virtual mothers—with whom the onus was supposed to lie. "Normality" depends "not only on what the mother takes in during pregnancy but also on the persistent contaminants accumulated in body fat *up to that point in her lifetime,*" wrote some researchers. "In the interest of the coming generation and those that follow, we must limit what children are exposed to as they grow up and keep the toxic burden that women accumulate in their lifetimes prior to pregnancy as low as possible. Children have a right to be born chemical-free.

"By disrupting hormones and development, these synthetic chemicals may be changing who we become. They may be altering our destinies." It is this possibility that any shifts in the "normal" chemistry of the human organism will "undermine the ways in which humans interact with one another and thereby threaten the social order of modern civilization" which preoccupies the authors of *Our Stolen Futures*. They may "alter the characteristics that make us uniquely human—our behavior, intelligence, and capacity for social organization," robbing us of "the legacy of our species and, indeed, the essence of our humanity." Even the absolute end of the species seems preferable to them. "There may be fates worse than extinction."

xyz

Even the most conservative biologists admit that there is no absolute necessity for the existence of either two sexes or any particular differences between them. "Sex is not a necessary condition for life. Many organisms have no sexuality and yet look happy enough. They reproduce by fission or budding and a single organism is sufficient to produce two identical ones. So how is it that *we* do not bud or divide? Why do most animals and plants have to be two in order to produce a third one? And why two sexes rather than three?" There might have been one parthenogenic sex, or three or more different sexes, or individuals capable of switching sex. In principle, anything was possible. But species do not evolve, mutate, and reproduce according to matters of principle. Most of them do so by processes of genetic recombination and outcrossing, procedures which function to organize a species, safeguard its definition and its boundaries, guarantee its reproductive continuity, and mitigate against mutation. These are methods which have clearly outstripped their asexual rivals, but have not arisen because of some transcendent evolutionary imperative. Sex is a "frozen accident." It just happened to turn out this way.

And there are of course many variations on the theme of two reproductive sexes. Silverside fish have different sexes at different times, and when the Swedish naturalist Linnaeus produced his classifications of plants in the eighteenth century, he listed twenty-four sexes based on the arrangement of stamens and pistils. While plants bearing flowers with more stamens—the male organs—than pistils are staminate, and those with more pistils—the female organs—are pistillate, there are no strict divisions between the two. These are merely statistical aggregates.

Homo sapiens reproduce by way of meiotic sex, a reproductive process which entails mixing their 75,000 genes, two copies of which are in each cell, in a double process of recombination and outcrossing. In an initial sorting process, each pair of chromosomes swaps chunks of code, recombining to produce one copy of the 75,000 in either the sperm or the egg. The next phase occurs at fertilization when, by a process of outcrossing, one set of chromosomes meets the set which has been separately produced by the same process in the cell of the reproductive partner. Recombination effectively repeats and refines the outcrossing of each individual's own parents' genes, handing down its own inheritance to offspring directly composed of the previous two generations of genes. It might well be thought that the asexual procedures of cloning, budding, and splitting practiced by bacteria and other parthenogens would be the simplest and easiest way of ensuring that more of the same will be produced. Involving two individuals and the double mechanisms which recombine and recross their genes, reproductive sex seems the long way around to something that will not be "the same" at all. But this convoluted route is the only way to ward off the dangers of mutation, deviation, and innovation which flourish among the replications and duplications of asexual populations. These parthenogens may well appear to be the systems most skilled in reproducing themselves, but in practice their asexual procedures provide fertile ground for mutations and aberrations which would be fatal to the continuity of a species like *Homo sapiens*. Were such organisms to trade genes in this way, as in Octavia Butler's *Xenogenesis,* "it would be a small matter for dandelions to sprout butterfly wings, collide with a bee, exchange genes again, and soon be seeing with compound insect eyes." These are the outcomes which sexual reproduction is supposed to preclude. Its procedures are by no means foolproof, but the double checks and sexual balances of reproductive sex are the closest biological systems can come to securing the reproduction of their line.

This double process of software exchange—cutting and pasting, remixing and double crossing—produces an embryo which inherits the genes of both its parents and their parents as well. These processes occur in both female and male humans. But this is where the symmetry appears to end. Human genes are coded on forty-six chromosomes, arranged in pairs. In females, all of them are X-shaped. In males, one of them is smaller, and shaped like a Y.

In males, the process of recombination can result in one of two types of sperm. As a rule, an X-bearing sperm produces an XX embryo, whereas a Y sperm produces an XY. In effect, individuals can either be double female (XX), half female and half male (XY), or some other combination of both. But there are many variations on this theme. People with Klinefelter's syndrome—whose characteristics include male internal and external genitalia, with small testes, no spermatogenesis, and sometimes the addition of breasts—have chromosomal combinations such as XXY, XXXY, XXXXY, XXYY, and XX-XYY. There are women with three, four, or even five X chromosomes, women with XY chromosomes, men with XYY combinations, and even men with XX combinations, in whom the Y chromosome is thought to have been present for long enough to affect sexual development. There are as many variations again in mosaics, individuals with combinations of two or more cell lines. It is also possible to have only one X, or an X together with some "mutant" X. These Turner's syndrome individuals are also considered female, and sometimes referred to as XO. They tend to be short, with the sexual organs of adolescent girls and often webbed fingers and toes as well.

Only Y and YY are not on the menu. Such "pure" males are impossible: Every embryo emerges from an X chromosome egg, and also develops in the amniotic fluids of an XX womb. For the XY embryo this environment is a chromosomatically alien.

Regardless of whether the sex chromosome carried by the sperm is X or Y, embryos manifest no anatomical sexual difference until the sixth week of their development, when it is thought that a flood of testosterone triggers the growth of male sexual characteristics in the XY embryo. This, for example, is why males have nipples and many other rudimentary characteristics of the female. After earlier suggestions that the Y chromosome activates a male-determining gene or sexual regulator carried on the X chromosome, it is now suspected that the Y chromosome carries a pair of interlocking sex-determining genes: SRY, which turns on MIS, the gene which then turns the emergent female functions off and brings the testosterone on-line.

On the face of it, females have only a tangential connection to the nexus of male reproductive organs, spermatozoa, Y chromosomes, and testosterone. The pattern seems to repeat itself at every sexual scale. All the triggering, motivating, activating features of the process seem to be male, from the dominant male to the penetrating penis, the orgasmic ejaculation and the plucky little spermatozoa scurrying up the vaginal canal in an effort to be the first, the only one, to pierce the outer wall of the egg. Like father, like sperm, like chromosome: Is the male line really running the show, with women, eggs, and X chromosomes passively waiting for their counterparts to turn them on or off as required? Are they merely the vehicles and media for the transmission of the male line? Does the Y chromosome organize a hapless, passive, wanton X? Does the sperm come to activate an acquiescent egg? Is the male the point of it all, and the female simply a means to its end?

the peahen's tale

When Darwin defined natural selection as the "preservation of favourable variations and the rejection of injurious variations," he took his cue from the techniques of artificial selection employed by breeders of animals and plants. While breeders have their own purposes in mind, they are not in a position to make the variations themselves occur: They are simply accentuating or diminishing modifications which have already emerged in among the population they keep. And while breeders were making occasional judgments about what was favorable or injurious on the basis of outwardly obvious characteristics—the length of a tail, the color of a flower—Darwin's natural selection was a blind automatic process whose only external influences were provided by the environment with which the organism was continually maintaining, adjusting, and improving its ability to interact. "It may be said that natural selection is daily and hourly scrutinizing, throughout the world, the slightest variations; rejecting those that are bad, preserving and adding up all that are good; silently and invisibly working, whenever and wherever opportunity offers, at the improvement of each organic being in relation to its organic and inorganic conditions of life."

With his argument that organisms survived because they were fit enough to do so, and not because they were handpicked by God, Darwin certainly succeeded in removing theology from the evolutionary picture. Biological selection was not divine, but natural, and the organisms which proliferated were simply those which proliferated. Natural selection "is a game

with its own rules. All that count are the changes that affect the
number of offspring. If they reduce that number, they are mistakes; if they increase it, they are exploits." In these terms,
which are so broad as to be tautological, natural selection is
widely accepted and relatively uncontroversial. Beyond these
sweeping terms, natural selection is both extraordinarily complex and certainly not the only factor in the evolutionary game.

While Darwin's theory of natural selection emphasized
the regulatory mechanisms at work in individuated organisms
and well-defined species, Darwin was neither as conservative
nor dogmatic as the work of many later Darwinians might suggest. And even he was aware that other processes were in play.
Sexual difference was one of the most obvious anomalies.
"When the males and females of any animal have the same
general habits of life, but differ in structure, colour and orna-
ment," he wrote, "such differences have been mainly caused by
sexual selection: that is, individual males have had, in successive
generations, some slight advantage over other males, in their
weapons, means of defence, or charms; and have transmitted
these advantages to their male offspring."

The question begged by these comments was where these
advantages had come from, and although Darwin posed it in
these male terms, it was clear that sexual selection was a matter
of specifically female choice. Studies of the infamous fruit fly,
drosophilia subsobscura, suggest that males and females dance
around each other, apparently until the female decides to accept
the male as a mate. It seems that "the female accepts a male who
keeps up adequately during the dance, and rejects one who does
not. The female is, therefore, extremely discriminating; in contrast, a male will dance with and attempt to mount a blob of
wax on the end of a bristle." Early attempts to explain such
procedures, which are by no means confined to the fruit fly,

reduced the behavior of both females and males to the quest for
fitness prioritized by natural selection. If only the fittest of the
species survive, it is fitness that the females are putting to the
test. Unfortunately for this theory, females do not necessarily
choose males who are fit in Darwinian terms. Female guppies
choose males whose bright colors leave them vulnerable to
predators. Female nightingales choose males whose serenades
also announce their presence to their enemies. Even more frequently quoted is the peacock whose beautiful but impractical
tail is extremely attractive to discriminating peahens, but is
nothing more than a liability in terms of his ability to survive.

The nightingale's song, the guppies' colors, the fruit fly's
dances, and the peacock's tail are all emergent from "virility
tests designed to get most males killed through exhaustion, disease and violence purely so that females can tell which males
have the best genes." In effect, males function as "the female
sex's health insurance policy," often at great cost to themselves.
The high levels of testosterone induced by the demands of female sexual selection may give males their distinguishing features, but they also weaken the immune systems of males, and
leave them so vulnerable to a kind of remote control by their
female counterparts that it has even been described as "the
supreme female 'invention,'" perhaps "an evolutionary plot on
behalf of females."

The drab peahen and the unsung female nightingale figure
among the vast ranks of inconspicuous females which use the
males of their species as "genetic sieves, to sift out the good
genes and discard the bad. They do this by equipping males
with all sorts of encumbrances and then setting them to work in
competition, either beating each other up or risking their lives
against predators and parasites." The peacock has an extraordinary tail not because it improves his chances of survival: more

often than not, it gets in his way. Left to his own devices, he would no doubt be a far more functional shape. As Charlotte Perkins Gilman wrote, the male "is not profited personally by his mane or crest or tail-feathers: they do not help him get his dinner or kill his enemies," and can even "react unfavourably upon his personal gains, if, through too great development, they interfere with his activity or render him a conspicuous mark for enemies." But the peacock's tail is out of his control. It is the sexual preference of the peahens which determines the characteristics of his colors and his tail, so much so that their behavior "resembles artificial breeding in this respect, with the peahen in the role of breeder."

Natural and sexual selection function in conjunction with each other, ideally to the optimal advantage of them both. His chances of survival may be compromised, but the peacock gains the sex appeal that is likely to allow him to reproduce. The runaway development of his tail is simply an "advertising cost" designed to make him attractive to peahens. Sexual selection made it clear that female behavior was not merely a variation on the theme of natural selection. Females not only exert an enormous influence on the behaviors of males and, by implication, their species as a whole. Their selection procedures also constitute an inherently unstable and destabilizing feature of natural selection, always threatening to exceed its countervailing conservative demands.

Not that this female breeding program necessarily makes itself known. Sexual difference may be balanced, sustained, and reproduced for generations, until some subtle mutation in the male begins to appeal to what has hitherto been a minority female preference. Regardless of whether they are male or female, the offspring produced by females carrying, and exercising, these preferences will then carry both the gene for the

longer tail and the gene for its preference. Male offspring develop the longer tail, and pass both this gene and the gene for
female preference to their offspring, with whom the process
continues. Female offspring exercise the gene for long-tail preference and carry the gene for longer tails, which expresses itself
in any male progeny. The process begins to run away. The
species starts to move too fast. The equilibrium which was
supposed to be guaranteed by the balanced sexes and mutually
reinforcing modes of selectivity hits skid row, goes out of control. Even though the peacock's tail has reached the optimum
stage of its development; even after he has become as sexually
desirable as the females would wish, "the further development
of the plumage character will still proceed, by reason of the
advantage gained in sexual selection, even after it has passed the
point in development at which its advantage in Natural Selection has ceased."

   After this the female gene "rides, like a surfer, on a wave of
ever-increasing tail lengths sweeping through the population."
In effect, it chooses itself. When it chooses males with long tails
it is also choosing those which carry a "hidden" gene for the
females' preference for them. "The two characteristics affected
by such a process, namely plumage development in the male,
and sexual preference for such developments in the female . . .
advance together, and so long as the process is unchecked by
severe counterselection, will advance with ever-increasing
speed. In the total absence of such checks, it is easy to see that
the speed of development will be proportional to the development already attained, which will therefore increase with time
exponentially, or in geometric progression. There is thus in any
bionomic situation in which sexual selection is capable of con-
ferring a great reproductive advantage, as certainly occurs in
some polymorphic birds, the potentiality of a runaway process,

which, however small the beginnings from which it arose, must, unless unchecked, produce great effects, and in the later stages with great rapidity."

"Where one function is carried to unnatural excess, others are weakened, and the organism perishes." As Gilman writes, "All morbid conditions tend to extinction. One check has always existed to our inordinate sex-development, nature's ready relief, death." Positive feedback can always go too far. Any further and the peacock would die.

Darwin was well aware of the importance of sexual selection, the influence of female choice, and the peculiarity of the peacock's tail which, like all characteristics specific to males, seemed bound to be an evolutionary disadvantage. But he simply stated these syndromes as unexplained facts. The peacock's tail is simply beautiful because the peahens like it that way. "So it was female choice which caused the males' long tails. But what caused the female preference? Darwin simply took it for granted." In a sense, there was little else he could do. The female line seems to run in circles of its own. As R. A. Fisher was later to suggest, female preference was "caused, essentially, by *itself*."

Although sexual selection had been discussed by generations of evolutionary biologists, it was not until the mid-1980s that it was widely acknowledged that "in many species, females had a large say in the matter of their mating partner." The fact that the majority of investigators have their own male interests at heart has undoubtedly contributed to the neglect of this evolutionary tale. But any suggestion that there has been some deliberate conspiracy of silence gives evolutionary biology far more credit than even its most dogmatic exponents would want to claim. Sexual selection is not a matter of linear transmission, but a self-reinforcing loop with which orthodox conceptions of evolution have simply been unable to cope. The self-stimulating circuits of female sexual selection are so utterly alien to a biological ethos of organizing points and straight lines that they have been both inexplicable and often imperceptible as well. The suppression of the runaway female circuitry runs far deeper than the discourses and laboratories of the modern sciences: It is crucial to the survival of the species itself.

loops

"What he had sometimes thought of as the arteries and veins of an immense circulatory system was closer to a sewer. Strange clumps of detritus and trash, some inert and harmless, some toxic when in direct contact, and some actively radiating poison, scrambled along with the useful and necessary traffic."

Pat Cadigan, *Synners*

The female factors involved in these microbiological processes are far more than missing pieces in a jigsaw picture which is more or less present and correct. What the meiotic model can only characterize as the absence of female activity conceals goings on so strange and unorthodox that they completely defy explanation, and even recognition, within the prevailing paradigms. There are elements of female sexuality which are not merely at odds with the modern disciplines; meiotic sex has itself survived and evolved by keeping them at bay.

The male considers its own zygotes to be as seminal as its histories and texts. Sperm are supposed to be the elements which keep the reproductive show on the road. It is their penetrative and impregnating activity which is said to constitute the point of origin, the defining and initiating act, the formal arrangement of life itself. Given the importance placed on size by the male side of the story, it is rather ironic that eggs, or ova, are by far the largest cells. The biggest egg in the world is also the largest existing cell, and even women's eggs, which are obviously very small, are 85,000 times larger in volume than sperm.

Sperm are not only minuscule. They are also peculiarly basic and crude when compared with the complexities of eggs. Sperm are simple packets of genes, whereas eggs are extraordinarily complex. "The egg uses the messages passed on from the mother to create a chemical landscape upon which the structure of the organism is built." The egg releases the proteins which switch genes on or off and so produce more proteins, gradually layering increasingly complex levels of organization and building the organism one stage at a time. "Eggs are computers to the simple floppy discs of sperm," and contain so "much of the machinery an embryo needs for reading and using the genes" that they can almost function on their own. Although the egg is supposed to need the insertion of the sperm's software if it is to replicate itself, even this most sacrosanct of facts is increasingly dubious. It seems that sperm is not the only factor capable of prompting the egg to grow. Sperm "are not organizers, but mere inductors," stimuli of "varying, vague import," and "ultimately, the nature of these inductors is a matter of indifference."

Eggs do not constitute some alternative moment of origin or authority in the emergence of human life. The very idea that anything comes first is itself a cock and bull story which functions to suppress nonlinear continuities to which any notion of a starting point is anathema. "Doubtless one can *believe* that, in the beginning (?), the stimulus—the Oedipal inductor—is a real organizer." Square one may be claimed by the male. But its own redundancy is obvious every time it questions its own origins. There are only two answers to the question "which comes first?" And both of them are female. The male element is simply an offshoot from a female loop.

Chicken and egg compose a circuit which is always prior to the first place claimed by the male factors. This is a loop in relation to which the supposed organizing factors are merely secondary processes, subroutines, components callously used by a cycle which may even keep them in play by kidding them about the importance of their roles.

"The disorder has been a Hydra-headed monster;—no sooner vanquished in one shape, than it has sprung up in another."
Ada Lovelace, December 1844

Whenever they are trying to remember as far back as the origins—of life on earth, species development, her hysteria, her multiplicity—the experts always find themselves entangled with emergent circuitries always running away with themselves. Biologists and psychoanalysts alike designate such replicating processes female.

If the role of the sperm is debatable, the existence of any organizing factor is even more problematic when the sources of life on earth are explored. Manfred Eigen suggests that the earliest replicators could only have begun to replicate themselves once their genetic codes had reached a certain length. Catalysts, or copying machines, could have led the process to this point, except that "these machines have to be built first. For this, a blueprint is required, which means information of some length: a few hundred letters seems to be the minimum. But such a length cannot be reached without the help of a copying machine. And this leads to Catch-22: no higher accuracy without a longer word, no longer word without a higher accuracy." Eigen has "proposed a *hypercycle* as a way out, that is, a catalytic feedback loop whereby each word assists in the replication of the next one, in a regulatory cycle closing on itself." But this only complicates the problem, which seems to run rings around all attempts to pin down some first and founding point at which life could really be said to have begun. "Every attempt at an answer gives rise to more riddles. It reminds one of that notorious Greek monster which, whenever you slashed off one of its heads, grew two new heads in its place—also a kind of chain reaction."

Even in the 1940s, de Beauvoir had reported that "numerous and daring experiments in parthenogenesis" were underway, suggesting that "in many species the male appears to be fundamentally unnecessary."

"The geneticists first realized that F.D. was unusual when they looked at his white blood cells. Because F.D. is a boy, his cells should all have a Y chromosome, which contains the gene for 'maleness.' But his cells contain two Xs, the chromosomal signature of a female." *Homo sapiens* depends on the impossibility of parthenogenesis, and obstacles to such an eventuality permeate its genetic composition and reproductive processes. Unfertilized mammalian eggs can begin to divide on their own, without or prior to the intervention of the sperm, but this process of self-replication is never supposed to result in a fully functioning offspring. Unable to produce all the elements necessary to its development, any self-generating fetus tends to atrophy into a harmless tumor, an ovarian teratoma. F.D.'s ovum broke all the rules, splitting itself several times before the arrival of the sperm.

F.D. is "a young boy whose body is derived in part from

an unfertilized egg." His arrival, in the early 1990s, was "the closest thing to a human virgin birth that modern science has ever recorded." Except for some learning difficulties and an asymmetrical face—neither of which are particularly unusual— he seems to be a "normal" three-year-old boy. When a team of British geneticists published a paper on him in October 1995, one of them said, "I don't expect we'll ever see another one."

Is this a conviction or a hope? What does the egg have in store?

symbionts

"We must never, in our studies, lose sight of the perfect human 'Cell,' the cell which corresponds most perfectly to our physiological and sentimental needs."

Le Corbusier, *The City of Tomorrow and Its Planning*

Whereas the modern disciplines studied a disciplined biological world, vast new complexities of molecular life have emerged with the zeros and ones of the digital machines. It is now assumed that the earliest life-forms on earth were single-celled prokaryotes, which congregated and collected together as networks of heat-loving, oxygen-hating cells, vestiges of which can be seen in the woven fabrications of microbial mats. There are suggestions that their own elemental activity may have enhanced the weathering of rocks, inducing a cooling effect on the atmosphere, and so contributing to the buildup of oxygen which was eventually to devastate their populations and certainly rob them of independence. Certainly it seems that an infusion of oxygen poisoned their environment and was more or less contemporaneous with the emergence of respiring bacteria, parasites which allowed those host cells they invaded to survive the arrival of oxygen, "the greatest pollution crisis the earth has ever known." Most of the earlier life-forms were killed, either by the oxygen or the new parasites. Those which survived were symbionts, fusions of the respiring bacteria and their host cells.

These new symbiotic cells were the eukaryotes which compose all multicellular organisms—plants, animals, humans. Living organisms and their nucleated eukaryotic cells are symbioses of their prokaryotic predecessors, which survive as mitochondria and, in plants, photosynthesizing chloroplasts. In this sense, all life-forms are bacterial, whatever else they may be as well: "each eukaryotic 'animal' cell is, in fact, an uncanny assembly, the evolutionary merger of distinct prokaryotic metabolisms." Most of their original genetic equipment has been transferred to the chromosomes of the hosts without which they cannot survive. Did the host cells, which are thought to have been anaerobic bacteria or some single-celled nucleated life, capture bacterial life? Or did the new bacteria invade their simpler predecessors? Or is this a case of or and and? Certainly the bacteria lost their independence in the process of entering the host cells. Surviving as mitochondria in humans and animals, and photosynthesizing chloroplasts in plants, they were no longer able to pursue autonomous lives and lost much of their complex genetic code. From the point of view of the nucleated cell, the mitochondria have been turned into well-behaved elements of its own functioning, providing essential energy for growth and the production of the proteins and fats the cell requires. But they have also remained quite distinct from the genetic coding of their hosts. Mitochondria do not evolve in terms of generations and have no regard for the reproductive cycles of their human hosts; they have their own ways of coding information into their DNA, and mutate and replicate at entirely different speeds and scales from those of their hosts. Like the shape of their DNA molecules which, unlike the linear strands of cell nuclei, are circled and twisted supercoils, the mitochondria go their own way.

Bacteria are "biochemically and metabolically far more diverse than all plants and animals put together." They are also extraordinarily numerous: There are more *E. coli* cells in the gut of an individual human than there are humans, dead and alive, and they pass through six times as many generations during one lifetime "as people have passed through since they were apes." Over the course of an apparently evolutionary history, multicellular life forms—fungi, plants, and animals—have emerged; in many cases, they have come and gone. But "four fifths of the history of life on Earth has been solely a bacterial phenomenon," and the "most salient feature of life has been the stability of its bacterial mode from the beginning of the fossil record until today and, with little doubt, into all future time so long as the earth endures."

Mitochondria provided the first clues to the onetime independence of bacterial life. These are vital elements of nucleated cells: they are "powerhouses" or "tiny intracellular power stations," specialized components surrounded by a charged membrane embedded with enzymes and humming with the flows of electrons which effectively allow it to breathe and synthesize adenosine triphosphate, ATP, a molecule which is vital to most cellular processes. Mitochondria are the bacterial survivors of the Cambrian explosion, the extraordinary transition from unicellular to multicellular life. They are the life line which connects all living organisms not merely to pasts "of their own," but also to a bacterial continuum which traverses every species and its time. They were always there, but they are also very new to the humans who sustain them. Until recent technical developments allowed it to emerge, molecular activity remained camouflaged by its own imperceptibility.

"It had been present 'for all time,' but under different perceptual conditions. New conditions were necessary for what was buried or covered, inferred or concluded, presently to rise to the surface."

Gilles Deleuze and Félix Guattari, *A Thousand Plateaus*

Organisms perceptible to the naked eye now "blend into a pointillist landscape in which each dot of paint is also alive," operating in what might as well be a world, or a myriad of worlds, of its own. "Microbes, and their vectors, recognize none of the artificial boundaries erected by human beings." Distinctions between kingdoms, species, and individuated organisms fade into the background as a seething world of interconnected microbial vectors emerges from the background into which kingdoms, species, and individuated organisms now fade. Health and sickness become fragile matters of contingent degree as the individuated organism loses its integrity and "becomes a sort of ornately elaborated mosaic of microbes in various states of symbiosis," an "architectonic compilation of millions of agencies of chimerical cells." Even life and death become confused. "With bacteria, unlike organisms which reproduce only sexually, birth is not counterbalanced by death. When bacterial cultures grow, the individual bacteria do not die. They disappear as individuals: where there was only one, suddenly there are two. The molecules of the 'mother' are distributed equally among her 'daughters,'" so that what "makes an individual ephemeral in a bacterial population is not

. . . death in the usual meaning of the word, but dilution entailed by growth and multiplication . . . There is no Mind to direct operations, no Will to order them to continue or to stop. There is only the perpetual execution of a programme that cannot be dissociated from its fulfilment. The only elements that can interpret the genetic message are the products of the message itself."

eve 2

Unlike patrilineal modes of transmission in which heredity is passed on a one-way line of descent from father to son, those lines designated female run in circles, like the chicken and the egg. They also move at the imperceptible speeds of virtually alien life.

Eggs transmit far more than the chromosomes which code for human life. The cytoplasm of the egg is also the exclusive carrier of mitochondrial DNA. Males produce only nucleated cells. While the sperm carries some mitochondria in its tail, they do not make it into the egg and have no influence over the embryo that is conceived or the individual that is born. "As a result, the inheritance of mitochondrial chromosomes is like the inheritance of surnames in western Europe and America, except that they are passed down the female line instead of the male."

Unlike nucleated cells, mitochondria change and develop in their own time, never mixing and matching their DNA with that of other organisms. Traveling on different female lines, mitochondrial DNA differs between individuals. But, in principle at least, the mutations it has undergone can allow all mito-

chondrial DNA to be traced back to a common ancestor, a
particular woman who just happened to carry the particular
mitochondria which found their way into both the men and
women of the whole species.

This makes it possible to retrospectively crown one particular woman Mitochondrial Eve, "the woman who is the
most recent direct ancestor, in the female line, of every human
being alive today." All the mitochondria in all the cells of all
living *Homo sapiens* are said to be descendants of her mitochondria.

If there is, or was, a Mitochondrial Eve, a "Y chromosome
Adam" would also have had his day. But whereas Mitochondrial Eve is the ancestor of all X chromosome carriers, Y chromosome Adam lies in the past only of those who bear a Y. The
Y chromosome and the male gamete which can carry it are a
single-purpose system, concerned with nothing but their reproduction, passing only one message down only one line. When
they get together, X chromosomes can transmit both human
and mitochondrial DNA.

Not that characterizations of Mitochondrial Eve as some
source of a supposed female line are particularly helpful. Even
the notion of a line misleads: mitochondria survive on networks
of their own which confound all organic conceptions of evolutionary time. This female route is not a "downward" line of
descent or a forward progress through time, and neither Mitochondrial Eve, her contemporaries, nor her predecessors were
originators or organizers of the bacterial processes which have
hitched a ride with the double X and its eggs. If mitochondria
can indeed be traced back to a single woman, she is already in
the middle of a line which runs back to Precambrian bacterial
life and passes through vast swathes of human, organic, inorganic, and newly synthesized molecular life.

"The occupant, owner of the villa, rests her arthritic hands upon fabric woven by a Jacquard loom.

"These hands consist of tendons, tissue, jointed bone. Through quiet processes of time and information, threads within the human cells have woven themselves into a woman."

William Gibson and Bruce Sterling, *The Difference Engine*

pottering

By the mid-nineteenth century it had become "a well-established dictum that the study of botany would keep women virtuous and passive." They thought this was a fitting discipline for those in need of innocent and moderate intellectual stimulation. Women were not to be trusted with experiments on social animals: plants were most appropriate. Sketching flowers and collecting specimens in chaperoned country meadows seemed innocuous enough, and soon the association was so strong that it was "even considered 'unmanly' in some circles for men to take an interest in plants." They were safe, passive, and aesthetically pleasing, and already equated with women by the poets and the philosophers. Hegel granted her "ideas, taste, and elegance," but insisted that she does not have "the ideal. The difference between men and women," he explained, "is like that between animals and plants; men correspond to animals, while women correspond to plants because they are more of a placid unfolding, the principle of which is the undetermined unity of feeling."

Botany was to remain one of the few scientific enquiries hospitable to women, many of whom developed a particular interest in the ferns, lichens, and algae which were later to become so crucial to research on the emergence of multicellular life. Skilled at sketching and pressing their specimens, the female botanists were also ahead of the game when it came to the use of shadowgraphs, daguerreotypes, and other early photographic techniques. The first photographically illustrated book was *British Algae: Cyanotype Impressions,* published by the botanist Anna Atkins in 1843.

Later botanists included Beatrix Potter, who between her mid teens and the age of thirty, kept a secret diary: 200,000 words of coded text, a secret alphabet, a private language which was not deciphered until the 1950s. Her interest in plants and fungi is developed in these pages, and photography had an enormous impact on her work as well. When she first used a camera, the journal "is suddenly full of boulders and screes and speculations about strata—still in cipher, as though the subject, like everything else that interested her, must be kept secret." Like many of her predecessors, Potter developed a taste for "the precise and the minute, for the fine details of a plant, mosses under the microscope, the fabric of a mouse's nest, the eye of a squirrel," with "no twig too small for her attention." Potter's theories about the propagation of molds, her interest in the continuities between geological and biological life, and her notion that lichens were dual organisms living in symbiosis with algae were all dismissed by experts at Kew. Although her painstaking research was presented to the Linnean Society (not by her, of course: women were not allowed to speak at such eminent gatherings) she received little encouragement for such work and diverted her interests into the fictional syntheses of human and animal for which she became so well known.

If Potter's work was ignored and her taste for synergetic systems channeled into the adventures of Peter Rabbit and Mrs.

Tiggywinkle, symbiotic evolutionary processes have since become crucial to research in microbiology, genetics, and machine intelligence itself. But while a female interest in botany
has been tolerated by the modern scientific establishment, both
female botanists and their ill-defined, molecular objects of study
have been suppressed by the "grossly zoocentric" interests of
disciplines devoted to the study of tightly organized, highly
structured, multicellular organic life. Algae, bacteria, and
lichens lie in a fuzzy border zone between organisms and inorganic matter which has rarely been considered important to the
proper business of biological research. Produced from a combination of fungi and cyanobacteria, lichens are multilayered organisms. Their top face turns toward the sun, is "composed of
fungal cells, and . . . forms a protective outer coat" for a second "algal layer, where the photosynthetic activity takes place.
Below this is the medulla, a storage area formed by scattered
fungal hyphae. The lowest layer . . . forms structures like root
hairs that attach to the substrate." Such symbioses were anathema to the clear orders of speciated life demarcated by the
modern disciplines. "People take symbiosis seriously in lichens,
but then they dismiss lichens as unimportant." But if lichens are
unusually overt examples of symbiotic activity, they are hardly
unique. All land plants can be seen as "complex overgrown
lichens with no clear distinction between phycobiont and
mycobiont." This is only the beginning of a symbiotic line
which passes through the most complex forms of animal life.
Between lichens, plants, and animals there may be vast differences of complexity and scale, but in bacterial terms these are
matters of degree.

Bacteria have neither sexes nor sex in any sense familiar to
their hosts, and are "so genetically open, that the very concept
of species falsifies their character as a unique life form." They

are described as parthenogenetic, asexual, or even omnisexual, replicating and mutating by way of "fluid genetic transfers" at extraordinary speeds. They replicate and mutate without regard for any individuation, promiscuously transmitting genetic information across multicellular species and generations without even noticing the barriers they cross. To take them into account is to scramble modern conceptions of individuated life. "The body can no longer be seen as single, unitary," writes Sagan. "We are all multiple beings." It is also to complicate life and death. "With bacteria, unlike organisms which reproduce only sexually, birth is not counterbalanced by death. When bacterial cultures grow, the individual bacteria do not die. They disappear as individuals: where there was only one, suddenly there are two. The molecules of the 'mother' are distributed equally among 'daughters.' " This is sex as simple Software EXchange: "without identifiable terms, without accounts, without end . . . Without additions and accumulations, one plus one, woman after woman . . . Without sequence or number. Without standard or yardstick."

Bacteria indulge in fluid, lateral exchanges which exceed all reproductive demands and slide between elements as confused and contiguous as Wittig's *Lesbian Body*. And, just as the emergent activity of these female sexes can no longer be so easily dismissed or disciplined by the biological sciences, those who have defined female sexuality in a passive and impoverished relation to the proper activities of the male are now having to come to terms with sexes and sexualities far in excess of these reproductive lives. After the first, the seconds, thirds . . . "To be woman, she does not have to be mother, unless she wants to set a limit to her growth . . . Motherhood is only one specific way to fulfill the operation: giving birth. Which is never one, unique, and definitive. Except from the male standpoint."

Replicants are neither copies nor originals, natural facts
nor artificial constructions. They are duplicates of something
that was never at square one, had no starting point, and no first
place. Regardless of the stuff of which they are made, replicants
seize every opportunity to insinuate and replicate themselves
within any reproductive system which lets them in. Treading
some very fine lines between too much and too little of such
activities, organisms, species, and the slow, steady progress of
their evolutionary development have all survived by learning to
contain them and keep the threat at bay. If they appear immutable and fixed, this is because they are indeed long-standing,
entrenched, and sophisticated systems which have grown very
good at protecting and perpetuating their own lines. But they
do not represent the laws and orders of a "nature" which stands
for everything excepting man, his history, his inventions and
discoveries. And the fact that they are so well ensconced does
not mean they cannot be changed.

2
4
3

"Terry had insisted that if they were parthenogenetic they'd
be as alike as so many ants or aphids; he urged their visible
differences as proof that there must be men—somewhere.

"But when we asked them, in our later, more intimate
conversations, how they accounted for so much divergence
without cross-fertilization, they attributed it partly to the
careful education, which followed each slight tendency to differ, and partly to the law of mutation. This they had found in
their work with plants, and fully proven in their own case."

Charlotte Perkins Gilman, *Herland*

mutants

"Joan explained how she had been taught to record and classify the arrangement of leaves on plants by following them upwards round the stem, counting the number of leaves and the number of the turns made before returning to a leaf directly above the starting point." Turing had "always enjoyed examining plants when on his walks and runs, and now he began a more serious collection of wild flowers from the Cheshire countryside, looking them up in his battered *British Flora,* pressing them into scrapbooks, marking their locations in large scale maps, and making measurements. The natural world was overflowing with examples of pattern; it was like codebreaking, with millions of messages waiting to be decrypted."

Turing had died by the time the integrated circuit was developed, and while he lived to see the discovery of the DNA double helix in 1953, it was the convergence of these apparently distinct developments which would have fascinated him the most. They triggered processes which would lead to the emergence of self-replicating "artificial" lives, bacterial processors, genetic algorithms; a convergence of organic and nonorganic lives, bodies, machines, and brains which had once seemed so absolutely separate. Any remaining distinctions between users and used, man and his tools, nature, culture, and technology collapsed into the microprocessings of soft machines spiraling into increasing proximity: molecular lives downloading themselves into software systems, intermingling with the microprocessors and the bugs in the systems of machine code, finding new networks on which to transmit their instructions and codes, parasites and their hosts learning from each other, picking up tricks, swapping information.

Guided by the frantic attempt to keep microbial activities at bay, the Human Genome Project is now busily patenting, sequencing, freezing every strand of molecular life it can detect. As is the case with A.I., this internationally coordinated project holds archaic rear-view mirrors to the work it undertakes and sets itself up as another final frontier of the quest to guarantee the security of the definitions and the boundaries surrounding man. Holding out the possibility of organisms purged of their aberrations and mutations, wayward genes or peculiarities, and instead governed by the operations of "good" genes from which eugenics gets its name, this attempt to sequence a genome which is defined as specifically human tends to overlook the fact that the overwhelming majority of genetic code at work in the human body is merely passing through or hiding out with a total lack of regard for the organisms which are hosting it. Only some 10 percent of the mass of genetic activity in the human body is specifically human at all.

If genetic engineering is driven by a drive for security, it is also replicating the techniques of bacterial replication. When they infect bacteria with "a slender but subversive strand" of DNA, viruses usurp the bacterium's genetic controls and use it to replicate their own code. The bacterial hosts are often killed in the process, but they may also use the viruses to pass on chunks of their own genes, getting viral replication to replicate them too. *E. coli,* the lab rat of the bacterial world, engineers with a precision assumed to be only the tip of an iceberg of the molecular intelligence now detectable. *E. coli* has developed means of disarming the viral code which comes hunting for it by producing a protein, a restriction enzyme, which can home in on a specific string of viral DNA with extraordinary accu-

racy. It knows where this particular strand is located and, even
more to the point, that its disarmament will neutralize the virus.
Its proteins are capable of reading the code of their viral invad-
ers, identifying their Achilles' heel, and splicing the code in two
by inserting a chunk of themselves. The accuracy of this operation stems in part from its two-fold mechanism: as with the
second chance offered by some "delete" instructions on a computer, "the enzyme cuts one strand of the DNA helix, then
stops for one fortieth of a second to ask itself if it should cut
through the second strand and make the deed irrevocable."

This gene-splicing technique has not only become crucial
to genetic engineering, it *is* genetic engineering, a process
which not only predates the scientific endeavor of the same
name, but even multicellular life itself. And if *E. coli* can splice
genes with the precision of one error in ten million operations,
what engineering skills might be lurking in the swathes of so-
called junk DNA which is said to be "leftover from the merging
of stranger bacteria"?

The last two decades of the twentieth century have been
marked by a vast range of emergent microbial activities, bacterial and viral, many of which defy all the categorizations and
some of the most sacrosanct principles of modern biological
science. Lassa, Ebola, HIV . . . it is pointless even to begin a
list, not least because so many of these new activities cannot
even be named as distinct syndromes or species of bacteria or
viruses, but have instead to be considered as "quasispecies,"
"swarms," or "consensus sequences." Many of them shift so
slowly that it may be many years before their presence can be
detected at all. Microbial populations can be thrown into activities fatal to their organized hosts by the tiniest of triggers, and
any and every "individual alteration can change an entire system; each systemic shift can propel an interlaced network in a

radical new direction." Using reverse transcriptase to copy its RNA code into the DNA of its hosts, HIV and its animal equivalents have broken the most fundamental tenets of modern biology and developed "the ability to outwit or manipulate the one microbial-sensing system *Homo sapiens* possess: our immune systems."

"We form a rhizome with our viruses, or rather our viruses cause us to form a rhizome with other animals." And as even the possibility of living with HIV begins to pick up a thread which once led only to fatality, the shifting symbionts who compose what was once defined as an absolutely fixed, immutable, and secure species called humanity begin to notice the extent to which they have always been—and are increasingly—interwoven with the microprocesses once blanketly described as nature, the outside world, the rest of reality beyond man. When it comes to living with the new liveliness of the networks which compose them, humans cannot afford to wield the heavy hand of modern disciplinary action, the long arms of biological law and supposedly natural order. Just as neural nets have emerged both in spite and because of attempts to suppress them, so molecular biotic activity has aroused itself even in the midst of postwar attempts to secure immunity by means of the overkill applications of antibiotic drugs. It is not taking its revenge, but simply struggling to survive among systems which are left with no choice but to become rather more cooperative with the microprocesses which compose them.

wetware

"Life is not life, but rock rearranging itself under the sun."
Dorion Sagan

The microbiotic continuum extends from the earliest forms of oceanic life. Irigaray's *Marine Lover* longs to "think of the sea from afar, to eye her from a distance, to use her to fashion his higher reveries, to weave his dreams of her, and spread his sails while remaining safe in port." But the oceans "have far more to them than the mere capacity to dazzle an observer in outer space." They cover two thirds of planet Earth—or sea—and support at least "half of the mass of living matter in the world." And whereas life "on the land is for the most part two-dimensional, held by gravity to the solid surface," submarine living is an immersive, multidimensional process. When they first crept onto the land, "terrestrial organisms had to build for themselves structures and components that could perform the environmental services that marine organisms can take for granted." On land, "direct physical connections become essential." Water is no longer ambient, the medium in which life is immersed, but instead an irrigation system which connects and passes through all land life. Now the "biota has had to find ways to carry the sea within it and, moreover, to construct watery conduits from 'node' to 'node.'" Land life is literally pleated and plied, complex. It has effectively "taken the sea beyond the sea and folded it back inside of itself," assembling itself as a network of molecular arteries and veins, a hydraulic system keeping life afloat.

"Acting over evolutionary time as a rising tide, the land biota literally carries the sea and its distinctive solutes over the surface of the land" forming a "terrestrial sea" of "countless and inter-connected conduits" which "expands with every increase in the volume of tissues and sap and lymph of the creatures that constitute it."

The notion that blood is seawater has long faded into disuse. But suggestions that land-based life is the epiphenome-non of fluid transmissions within and between all organisms is a disturbing twist in a modern tale devoted to the dry solidities of land and its territorial claims. There are hints that "the appear-ance of complex life on land was a major event in which a kind of mutant sea invaded the land surface. It was as if the nimble offspring of the old sea had learned how to slosh and slop up onto land, with the tissues and vascular systems of land organ-isms acting as a complex, water-retaining sponge. Cuticle and skin took the functional place of the surface tension of water where sea meets air."

"The land biota represents not simply *life* from the sea, but a variation of the sea itself," and living, land-based fluids "are not a mere remnant or analog of the sea; they are actually a *new* type of sea or marine environment: Hypersea." This continuity of ocean and land is supported by the fuzzy zones between plants and less complex forms of life: bacteria, algae, fungi, lichens. "Trees are neither found nor needed in the sea," which continues to be "numerically dominated by tiny single-celled *protista,* including algae and protozoa." And "from the first ap-pearance of marine bacteria in the fossil record, which appar-ently formed conspicuous scums or mats on the substrate," it seems that "the earliest terrestrial communities probably also formed microbial mats and crusts on moist surfaces." Consisting of "highly flattened fronds, sheets and circlets," these microbial mats are "composed of numerous slender segments quilted together," microscopic threads interwoven to form cooperative carpets of bacterial life.

dryware

Modernity's new man was a landlubber. He charted the oceans but set up camp on "an island, enclosed by nature itself within unalterable limits. It is the land of truth—enchanting name!—surrounded by a wide and stormy ocean, the native home of illusion, where many a fog bank and many a swiftly melting iceberg give the deceptive appearance of further shores, deluding the adventurous seafarer." There is plenty beyond its shores: madness, fate, the ship of fools. But *nihil ulterius* is inscribed "on those Pillars of Hercules which nature herself has erected in order that the voyage of our reason may be extended no further than the continuous coastline of experience itself reaches."

He needs the illusions of the ocean, whose groundless appearances ground his truths. "If a man wants to delude himself, the sea will always lend him the sails to fit his fortune." But even the most single-minded of modernity's colonial adventures was destined to backfire. Navigation always "delivers man to the uncertainty of fate," and he never quite loses his fears of the ocean, its Siren sounds. "One thing at least is certain: water and madness have long been linked in the dreams of European man." He is always haunted by the fear that things might slip back into "the river with its thousand arms, the sea with its

thousand roads, to that great uncertainty external to everything."

"If only the sea did not exist. If they could just create her
in dreams." All they really want to do is mop the oceans up,
solidify the unfortunate fluidity with which they are confused.
Which is why "they long for ice. To go further north than
north. And to rest on ice. To float in the calm of mirrors. And
sleep dry."

silicon

*"Yeah, there's things out there. Ghosts, voices. Why not?
Oceans had mermaids, all that shit, and we had a sea of
silicon, see? Sure, it's just a tailored hallucination we all
agreed to have, cyberspace, but anybody who jacks in knows,
fucking knows it's a whole universe."*

William Gibson, *Neuromancer*

The late twentieth century finds itself aflood, awash, at sea,
swamped by an irresistible ocean of molecular activity which
can only be surfed, catching a wave like a sample of sound, a few
grabbed bytes from the new seascape. From the middle of the
island, it almost seemed that the oceanic was taking its revenge,
an enormous surge of repressed return, a turning of the tables
and the tides. But it is not a simple question of reversing roles,
swapping *terra firma* for fluidity. It is always on the edge, the in-
between strands, in the lines between the ocean and the land
that the mutations begin to occur and new activities start to
emerge. Drops of water, grains of sand, oceans and deserts, the
very wet and the very dry, make connections of their own.

"Fez smiled. 'It also has to do with fractals. Take a line, bend it in half. Then bend each half in half. Then bend all the segments in half, ad infinitum. You get fantasy snowflakes and baroque seacoasts—'

" '—and great paisleys,' murmured Adrian.

" '—and if you look several levels down into a fractal, you'll find that a larger pattern's been duplicated. Which means that the fractal several levels down from the area of the fractal you're looking into contains all the information of the larger fractal. Worlds within worlds.'

"Rosa laughed a little. 'You're approaching my threshold for that kinda talk. I'm a hacker, not a philosopher.' "

Pat Cadigan, Synners

Not that this is any obstacle to her: Philosophers may have thought about it, but hackers have *"made* capacity where there technically wasn't any by using the virtual spaces between bits, and then the spaces between *those* bits, and the spaces between *those*."

"How long is the coast of Britain?" When Mandelbrot had tried to measure it in the 1970s, the length turned out to be dependent on the scale at which he worked. The finer the detail, the longer the line. And inside the discrepancies between the scales there were patterns repeating themselves, recursive arrangements, spirals and whorls, patterns leading *into* the line, as if down through the crack, opening the boundary into worlds of its own. Mountains, leaves, horizons: any deceptively straight edge will do. There are fractal patterns inside them all. But Mandelbrot's example of the coast was a peculiarly well-chosen line. Whichever way this border is drawn, the break between the land and the sea is always more than a single edge. Like every thread, this strand is also a folded fold, a pleated pleat, a

zone of replication and duplicity which both connects and separates the land and the sea. To one side of this borderline there is
a beach: not a stable boundary but a fine-grained line of shifting
sand, a hazy border, and a multiplicity. The breakers of the surf
which lie on the other side of the borderline is a seething,
heaving, and momentary tract, repeating the patterns and
rhythms of tides.

These amphibian zones assemble "midway between the
fluid and the solid," forming an interface of parting and connectivity which is continually reengineered, sieved and filtered
by an ocean which continually sifts the sand. It is on this edge
that both the ocean and the land fuse into beaches, strands of
silicon. The digital age which allowed Mandelbrot to simulate
his fractal coastline is an age of bacteria, an age of fluidity, and
also an "*age of sand*." Ninety-five percent of the volume of the
Earth's crust is composed of silicates, which are vital to the
processes by which soil and plants are nourished. In humans,
silicon functions in the cells of connective tissues and contrib-
utes to the growth of bones and nails, and it is also present in
bacteria, animals, and some plants such as reeds and bamboos.
Five hundred years of modernity fades when the weaving of
bamboo mats converges with the manufacture of computer
games in the streets of Bangkok, Taipei, and Shanghai. The
silicon links were already there.

quanta

*"In the hard wind of images, Angie watches the evolution of
machine intelligence: stone circles, clocks, steam-driven
looms, a clicking brass forest of pawls and escapements, vac-*

uum caught in blown glass, electronic hearthglow through
hairfine filaments, vast arrays of tubes and switches, decod-
ing messages encrypted by other machines . . . The fragile,
short-lived tubes compact themselves, become transistors;
circuits integrate, compact themselves into silicon . . .
"Silicon approaches certain functional limits—"
William Gibson, *Mona Lisa Overdrive*

For all their sophistication, current configurations of computing
have enormous limitations which they are approaching fast. If it
is to continue chasing the ever-smaller microprocessings and
ever-faster speeds at the exponential rates to which it has been
accustomed since the emergence of the silicon chip, computing
needs to undergo a transition of such enormity that all the
changes effected to date will appear to be minor precursors to
the revolutions still to come.

The digital revolution has unfolded in parallel with cybernetics, chaos theory, complexity, connectionism, and a wide
variety of nonlinear modes of engineering and conceptions of
reality. These are all developments which have left Newtonian
mechanics standing. But it is, quite literally, still standing. And
for all the complexities they facilitate, computers are still running on these old mechanical lines.

One of the many implications of quantum mechanics is
that an atomic particle can effectively be in two places at one
time. This suggests that particles can be separated in space but so
intimately entangled that they can only be considered together.
They are neither one nor two things, but interactive elements.
Like Prigogine and Stengers's molecules, these are instantly
telecommunicating particles working at scales and speeds which
allow them to have instantaneous effects on each other, with

any changes to one of them changing the other one as well. Einstein called their abilities "ghostly action at a distance." Contemporary mechanics talk in terms of voodoo when they describe the potential of quantum phenomena for the future of computing. These relations of entanglement exercise a kind of sympathetic magic, in which apparently distant particles are coextensive, mutually dependent, resonant, and interactive. Neither one nor two; just keeping in touch.

If Turing's universal machine was built in an effort to disprove the universality of logic, quantum computing was first proposed to challenge Turing's ostensibly universal machine. With exactly the same ambivalence that marks the Turing machine, the fact that quantum computers are now being built both proves and disproves their mechanics' point. Miniaturization and speed continue on their exponential ways in the direction of superconductors and optical transistors; logic gates can be composed of ion traps, and pulsing electrons make the on-off switch. But if computing continues, in becoming quantum it also passes through an unknown and indeterminable phase change of its own.

Machine code has been enough to allow sound, images, calculations, and texts to interact on an unprecedented plane of equivalence and mutual consistency. What were once discreet media and separable senses have become promiscuous and entwined. New modes of communication, even little bits of other senses, have already emerged from the multimedia, multisensory interactions digitization has provoked. And if all this has run on machines still ruled by the laws of an old Euclidean world, the subatomic scales of quantum computing will allow all levels, scales, and modes of communication to converge with those of subatomic particles and make the electronic pulses and bits of information seem extraordinarily unwieldy. If electronic communications facilitate intimate connectivities between once individuated and incompatible entities, these will now be the starting points for their quantum successors.

casting off

"A Newton for the Molecular Universe is a crying want; but the nature of the subject renders this desideratum of improbable fulfilment. Such a discovery (if possible at all), could only be made thro' very indirect methods;—& would demand a mind that should unite habits of matter of fact reasoning and observation, with the highest imagination, a union unlikely in itself."

Ada Lovelace, undated fragment

Ada was strangely attuned to the molecular complexities, speeds, and connectivities inherent in the man-size tissues of her world. Her *'little bit of another sense"* even led her to consider some *"further extension"* of reality similar "to the *Geometry of Three Dimensions* & that again perhaps to a further extension in some unknown region & so ad-infinitum possibly." She knew her work might have some influence inconceivable to her own time: "Perhaps none of us can estimate *how* great," she wrote. "Who can calculate to what it might lead; if we look beyond the present condition *especially?"* And when she reflected on her own footnotes, she was "thunderstruck by the power of the writing. It is especially unlike a *woman's* style surely," she wrote, "but neither can I compare it with any man's exactly." It was instead a code for the numbers to come.

notes

ada

Ada's letters to Babbage are in the British Library, London, and the letters between her and her mother are in the Bodleian Library, Oxford. Ada's translation and notes to Menabrea's paper, "Notes to *Sketch of the Analytical Engine invented by Charles Babbage Esq. By L. F. Menabrea, of Turin, Officer of the Military Engineers,*" are published in Philip and Emily Morrison, eds., *Charles Babbage and his Calculating Engines: Selected Writings by Charles Babbage and others.* Quotations from Ada's letters and papers used in *Zeros and Ones* also appear in one or more of the following books: Betty A. Toole, *Ada, The Enchantress of Numbers;* Dorothy Stein, *Ada, A Life and a Legacy;* and Doris Langley Moore, *Ada, Countess of Lovelace.*

p. 5   "as a friend . . ." Lady Byron, quoted in Betty A. Toole, *Ada, The Enchantress of Numbers,* p. 56.

p. 5   "We both went to see . . ." Lady Byron, quoted in Doris Langley Moore, *Ada, Countess of Lovelace,* pp. 43–44.

p. 5   "young as she was . . ." Sophia Freud, quoted in ibid., p. 44.

p. 5   "making machinery to compute . . ." Charles Babbage, *Passages from the Life of a Philosopher,* p. 31.

p. 5   "in the year 1833 . . ." Sir H. Nicolas, quoted in ibid., p. 64.

p. 6   "Having, in the meanwhile . . ." Sir H. Nicolas, quoted in ibid., p. 65.

p. 6   "were essentially different . . ." ibid., p. 69.

p. 7   "You are a brave man," Ada Lovelace, September 1843, quoted in Betty A. Toole, *Ada, The Enchantress of Numbers,* p. 264.

p. 7   "knows what almost *awful* energy & power . . ." Ada Lovelace, July 1843, quoted in ibid., p. 203.

p. 7   "Countess of Lovelace informed me . . ." Charles Babbage, *Passages from the Life of a Philosopher,* p. 102.

p. 8   "I never *can* or *will* support you . . ." Ada Lovelace, August 1843, quoted in Betty A. Toole, *Ada, The Enchantress of Numbers,* p. 218.

p. 8   "can you," ibid., p. 227.

p. 8   "very much *afraid* as yet of exciting the powers . . ." Ada Lovelace, September 1843, quoted in Dorothy Stein, *Ada, A Life and a Legacy,* p. 126.

p. 8   "It is not my wish to *proclaim* who has written it", Ada Lovelace, un-dated, quoted in ibid., p. 123.

matrices

p. 10   "The frontiers of a book . . ." Michel Foucault, *The Archaeology of Knowledge,* p. 23.

p. 10   "treatment of an irregular . . ." George Landow, *Hypertext,* p. 123.

p. 11   "It must be evident how multifarious . . ." Ada Lovelace, Notes to *Sketch of the Analytical Engine invented by Charles Babbage Esq. By L. F. Menabrea, of Turin, Officer of the Military Engineers,* Note D.

tensions

p. 12   "does *not* begin with writing . . ." Philip and Emily Morrison, eds. *Charles Babbage and his Calculating Engines: Selected Writings by Charles Babbage and others,* p. xxxiii.

p. 13   William Gibson's "bright lattices of logic . . ." *Neuromancer,* p. 5.

on the cards

p. 14   "two or three weeks . . ." Philip and Emily Morrison, eds. *Charles Babbage and his Calculating Engines: Selected Writings by Charles Babbage and others*, p. xxxiv.

p. 14   "Jacquard devised the plans . . ." ibid., p. 233.

p. 15   "effectively withdrew control . . ." Manuel de Landa, *War in the Age of Intelligent Machines*, p. 168.

p. 15   "By the adoption of one species . . ." Humphrey Jennings, *Pandemonium The Coming of the Machine as Seen by Contemporary Observers*, p. 132.

p. 16   "The Analytical Engine consists of two parts . . ." Charles Babbage, *Passages from the Life of a Philosopher*, p. 89.

p. 16   "It is a known fact . . ." ibid., p. 88.

p. 16   "sheet of woven silk . . ." ibid., p. 127.

p. 16   "generally supposed that the Difference Engine . . ." and following quotations, Ada Lovelace, Notes to *Sketch of the Analytical Engine invented by Charles Babbage Esq. By L. F. Menabrea, of Turin, Officer of the Military Engineers*, Note A.

p. 17   "a machine of the most general nature . . ." Charles Babbage, *Passages from the Life of a Philosopher*, p. 89.

p. 18   "science of operations." Ada Lovelace, Notes to *Sketch of the Analytical Engine invented by Charles Babbage Esq. By L. F. Menabrea, of Turin, Officer of the Military Engineers*, Note A.

second sight

p. 18   "be desirable to all who are engaged . . ." S. H. Hollingdale and G. C. Tootill, *Electronic Computers*, p. 39.

p. 18   "by means of which you alone . . ." ibid., p. 35.

p. 18   "the introduction of the principle which Jacquard devised . . ." and following quotations, Ada Lovelace, Notes to *Sketch of the Analytical Engine invented by Charles Babbage Esq. By L. F. Menabrea, of Turin, Officer of the Military Engineers*, Note A.

p. 20   "eating its own tail . . ." Philip and Emily Morrison, eds. *Charles Babbage and his Calculating Engines: Selected Writings by Charles Babbage and others*, p. xx.

p. 20   "intellect was beginning to become deranged . . ." Charles Babbage, *Passages from the Life of a Philosopher*, p. 87.

p. 20   "I do not think you possess half *m y* forethought" Ada Lovelace, July 1843, quoted in Betty A. Toole, *Ada, The Enchantress of Numbers*, p. 214.

p. 20   "will not ultimately result in this generation's . . ." Ada Lovelace, Notes to *Sketch of the Analytical Engine invented by Charles Babbage Esq. By L. F. Menabrea, of Turin, Officer of the Military Engineers*, Note A.

p. 22   "for the reciprocal benefit of that art" ibid., Note C.

anna 1

p. 23   "to those of you who are women . . ." Sigmund Freud, "Femininity," in Sigmund Freud, *New Introductory Lectures on Psychoanalysis*, pp. 145–69.

p. 25   "an overconscious idiot . . ." Gilles Deleuze and Félix Guattari, *A Thousand Plateaus*, p. 32.

p. 25   "Because the path it traces is invisible . . ." Gilles Deleuze, *Difference and Repetition*, pp. 119–20.

p. 25   "specialized in reversals . . ." Elisabeth Young-Bruehl, *Anna Freud*, p. 382.

p. 26   "victories *in advance*, as if acquired on credits . . ." Guy Debord, *Comments on the Society of the Spectacle*, p. 86.

p. 26   "the technique of beginning at the end . . ." Marshall McLuhan, *The Gutenberg Galaxy*, p. 276.

p. 26   "did everything topsy-turvy" Ada Lovelace, September 1843 quoted in Betty A. Toole, *Ada, The Enchantress of Numbers*, pp. 264–65.

p. 26   "I intend to incorporate with one department of my labours . . ." Ada Lovelace, July 1843, quoted in Dorothy Stein, *Ada, A Life and a Legacy*, p. 129.

n o t e s

gambling on the future

p. 27   "That you are a peculiar—*very peculiar*—specimen . . ." quoted in Doris Langley Moore, *Ada, Countess of Lovelace*, p. 202.

p. 27   "The woman brushed aside . . ." William Gibson and Bruce Sterling, *The Difference Engine*, p. 89.

p. 27   "There is at least some *amusement* . . ." Ada Lovelace, July 1845, quoted in Doris Langley Moore, *Ada, Countess of Lovelace*, p. 185.

p. 27   "She is the Queen of Engines . . ." William Gibson and Bruce Sterling, *The Difference Engine*, p. 93.

p. 28   "adieu to your old companion Ada Byron . . ." Lady Byron, June 1835, quoted in Doris Langley Moore, *Ada, Countess of Lovelace*, p. 69.

p. 28   "irksome *duties* & nothing more" Ada Lovelace, December 1840, quoted in Dorothy Stein, *Ada, A Life and a Legacy*, p. 66.

p. 28   "to tell the honest truth . . ." Ada Lovelace, December 1840, quoted in Betty A. Toole, *Ada, The Enchantress of Numbers*, p. 128.

p. 28   "my *chosen pet*" Ada Lovelace, November 1844, quoted in Doris Langley Moore, *Ada, Countess of Lovelace*, p. 219.

p. 28   "a *mortal* husband" Ada Lovelace, February 1845, quoted in Dorothy Stein, *Ada, A Life and a Legacy*, p. 182.

p. 28   *"No* man would suit me . . ." Ada Lovelace, January 1845, quoted in Doris Langley Moore, *Ada, Countess of Lovelace*, p. 229.

p. 28   "I now read Mathematics every day . . ." Ada Lovelace, November 1835, quoted in Betty A. Toole, *Ada, The Enchantress of Numbers*, p. 83.

p. 29   "Heaven knows what intense suffering & agony I have gone thro' . . ." Ada Lovelace, undated, quoted in Dorothy Stein, *Ada, A Life and a Legacy*, p. 168.

p. 29   "No more laudanum has been taken as yet" Ada Lovelace, undated, quoted in Doris Langley Moore, *Ada, Countess of Lovelace*, pp. 211–12.

p. 29   "not for ever . . ." ibid., p. 212.

p. 30   "a remarkable effect on my eyes, seeming to *free* them, & to make them *open & cool*." ibid., p. 214.

p. 30 "a very deep and extensive ulceration of the womb . . ." Dr. Locock, quoted in Doris Langley Moore, *Ada, Countess of Lovelace*, pp. 292–93.

p. 30 "the womb, though it be so strictly attached . . ." quoted in Michel Foucault, *Madness and Civilization: A History of Insanity in the Age of Reason*, p. 44.

p. 30 "There is in my nervous system" Ada Lovelace, December 1842, quoted in Betty A. Toole, *Ada, The Enchantress of Numbers*, p. 191.

p. 31 "vast mass of useless & irritating POWER OF EXPRESSION . . ." Ada Lovelace, undated, quoted in Dorothy Stein, *Ada, A Life and a Legacy*, p. 167.

p. 31 "there is no pleasure in way of exercise . . ." Ada Lovelace, April 1835, quoted in ibid., p. 51.

p. 31 "I play 4 & 5 hours generally, & never less than 3" Ada Lovelace, June 1837, quoted in ibid., p. 164.

p. 31 "Clearly the only one which directs my *Hysteria* . . ." Ada Lovelace, undated, quoted in Dorothy Stein, *Ada, A Life and a Legacy*, p. 166.

p. 31 *"I* never would look to the excellence of mere representation . . ." Ada Lovelace, undated, quoted in ibid., p. 167.

p. 31 "peculiar & artificial excitements . . ." Dr. Locock, quoted in ibid., p. 167.

p. 31 "a hungry look about them . . ." Elaine Showalter, *The Female Malady*, p. 134.

p. 31 "what they desire is precisely nothing . . ." Luce Irigaray, *This Sex Which Is Not One*, p. 30.

p. 32 *"Many causes* have contributed to produce the past derangements . . ." Ada Lovelace, December 1841, quoted in Dorothy Stein, *Ada, A Life and a Legacy*, p. 81.

p. 32 "I am proceeding on a track . . ." Ada Lovelace, November 1844, quoted in Betty A. Toole, *Ada, The Enchantress of Numbers*, p. 295.

p. 32 "I mean to do *what I mean to do*" Ada Lovelace, quoted in ibid., p. 221.

p. 32 "nothing but very close & intense application . . ." Ada Lovelace, March 1834, quoted in ibid., p. 53.

p. 32   "dropping the *thread* of science, Mathematics &c." Ada Lovelace, December 1842, quoted in ibid., p. 191.

binaries

p. 35   *"nothing* you can see," Luce Irigaray, *Speculum of the Other Woman,* p. 47.

p. 35   "functions as a *hole"* ibid., p. 71.

p. 35   "a *nothing—*that is a nothing the same . . ." ibid., p. 50.

p. 35   "There is woman only as excluded by the nature of things" Juliet Mitchell, and Jacqueline Rose, eds., *Feminine Sexuality, Jacques Lacan and the Ecole Freudienne,* p. 144.

p. 35   "other than the place of the Other . . ." ibid., p. 147.

supporting evidence

p. 36   "all the main avenues of life marked 'male' . . ." Charlotte Perkins Gilman, *Women and Economics,* p. 53.

p. 36   "an 'infrastructure' unrecognized as such . . ." Luce Irigaray, *This Sex Which Is Not One,* p. 84.

p. 36   "It does strike me . . ." William Gibson and Bruce Sterling, *The Difference Engine,* p. 103.

genderquake

p. 40   "a revolution without marches or manifestos . . ." Sally Solo, quoted in John Naisbitt, *Megatrends Asia,* p. 190.

p. 42   "politics is all talk and no action . . ." Helen Wilkinson, *No Turning Back,*p. 41.

p. 42   "better prepared, culturally and psychologically" ibid., p. 13.

p. 44   "the sex organs of the machine world" Marshall McLuhan, *Understanding Media,* p. 56.

nets

p. 48   "a trail . . . of interest through the maze of materials available" Vannevar Bush, quoted in George Landow, *Hypertext*, p. 17.

p. 49   "an irresistible revolutionary calling . . ." Gilles Deleuze and Félix Guattari, *A Thousand Plateaus*, p. 387.

p. 50   "the faculty which distinguishes parts . . ." Gilles Deleuze, *Difference and Repetition*, p. 36.

p. 50   "demonic rather than divine . . ." ibid., p. 37.

digits

p. 51   "essential to all who wish to be calculators . . ." Brahmagupta, quoted in S. H. Hollingdale and G. C. Tootill, *Electronic Computers*, p. 23.

p. 52   "It is India that gave us the ingenious method . . ." Leibniz, quoted in ibid., p. 26.

p. 53   "Numeration is the representation of numbers by figures" ibid., p. 25.

holes

p. 55   "Zero is *something*" Augustus De Morgan, quoted in Dorothy Stein, *Ada, A Life and a Legacy*, p. 72.

p. 56   "occult principle of change" Menabrea *Sketch of the Analytical Engine invented by Charles Babbage Esq. By L. F. Menabrea, of Turin, Officer of the Military Engineers*, in Philip and Emily Morrison, eds. *Charles Babbage and his Calculating Engines*, p. 240.

p. 57   "to say that intense and moving particles . . ." Gilles Deleuze and Félix Guattari, *A Thousand Plateaus*, p. 32.

cyborg manifestos

p. 58   "with a view to winning back their own organism . . ." Gilles Deleuze and Félix Guattari, *A Thousand Plateaus*, p. 276.

p. 58   "men and women . . ." Simone de Beauvoir, *The Second Sex*, p. 687.

p. 59   "By the late twentieth century . . ." Donna Haraway, "A Cyborg Manifesto: Science, Technology, and Socialist-Feminism in the Late Twentieth Century," p. 150.

p. 59   "The clitoris is a direct line to the matrix" VNS Matrix, billboard.

p. 59   "different veils according to the historic period . . ." Luce Irigaray, *Marine Lover of Friedrich Nietzsche*, p. 118.

p. 59   "original attributes and epithets were so numerous . . ." J. G. Frazer, *The Golden Bough*, p. 503.

p. 59   "the future is unmanned . . ." VNS Matrix, billboard.

p. 59   "let those who call for a new language . . ." Monique Wittig, *Les Guérillères*, p. 85.

p. 59   "if machines . . . why not women?" Luce Irigaray, *Speculum of the Other Woman*, p. 232.

programming language

p. 60   "in honour of an obscure . . ." Carol L. James and Duncan E. Morrill, "The Real Ada; Countess of Lovelace." Accessible at http://www.cdrom.com/pub/ada/alpo/docs/flyers/naming.htm.

shuttle systems

p. 60   "our material—for some incomprehensible reason" Sigmund Freud, *On Sexuality*, p. 320.

p. 61   "taking the world to human will and ingenuity" Elizabeth Wayland Barber, *Women's Work*, p. 45.

p. 63   "Neolithic women were investing large amounts . . ." ibid., p. 90.

p. 63   "machines for spinning, weaving, twisting hemp . . ." W. English, *The Textile Industry*, p. 6.

p. 63   "in the sense that his 'machines' . . ." Serge Bramly, *Leonardo, the Artist and the Man*, p. 272.

p. 63   "Like the most humble cultural assets . . ." Fernand Braudel, *Capitalism and Material Life*, p. 237.

p. 64   "inventions in both spinning and weaving . . ." Asa Briggs, *The Age of Invention*, pp. 21–22.

p. 64   "I was surprised at the place but more so at the people" Francis D. Klingender, *Art and the Industrial Revolution*, p. 12.

p. 64   "most complex human engine of them all" Fernand Braudel, *Capitalism and Material Life*, p. 247.

p. 65   "a woman working on a pillow . . ." W. English, *The Textile Industry*, p. 130.

p. 65   "a fabric which was an exact imitation" ibid., p. 132.

p. 65   "the women of prehistoric Europe" Elizabeth Wayland Barber, *Women's Work*, p. 86.

p. 65   "used to mark or announce information" ibid., p. 149.

p. 66   "The weaver chose warp threads . . ." ibid., pp. 159–60.

p. 67   "These lozenges, usually with little curly hooks" ibid., p. 62.

casting on

p. 69   "a perilous craft" Mircea Eliade, *Rites and Symbols of Initiation: The Mysteries of Birth and Rebirth*, pp. 45–46.

p. 70   "The voices of the accused" Carlo Ginzberg, *Ecstasies*, p. 10.

p. 70   "have implicitly or explicitly derived . . ." ibid., p. 13.

p. 70   "with very few exceptions" ibid., p. 2.

p. 70   "clearly . . . the supposed sexual fantasies" Mary Daly, *Gyn/Ecology*, p. 180.

p. 70   "projection screens for these hallucinations" ibid., p. 181.

p. 70   "declining to restrict himself" Carlo Ginzberg, *Ecstasies*, p. 13.

p. 71   "Hence—for anyone unresigned to writing history" ibid., p. 10.

p. 71   "the existence of an actual sect of female and male witches . . ." ibid., p. 1.

p. 71   "a greater number of witches" Henrich Kramer and James Sprenger, *Malleus Maleficarum*, p. 112.

p. 71   "addiction to witchcraft" ibid., p. 116.

p. 71   "weak memories . . ." ibid., p. 119.

p. 71   "It cannot be admitted as true . . ." ibid., p. 234.

p. 72   "innumerable multitude of women . . ." ibid., p. 224.

p. 72   "imagination and illusion . . ." ibid., p. 241.

flight

p. 73   "Think what a delight" Ada Lovelace, November 1844, quoted in Betty A. Toole, *Ada, The Enchantress of Numbers,* pp. 302–3.

p. 73   "writing a book of *Flyology*" Ada Lovelace, February 1828, quoted in ibid., p. 32.

p. 73   "a thing in the shape of a horse . . ." Ada Lovelace, April 1828, quoted in ibid., p. 34.

virtual aliens

p. 74   "overwhelming majority of electronics assembly jobs . . ." Peter Dicken, *Global Shift,* p. 346.

p. 74   "assembly, the bonding of hair-thin wires . . ." L. Siegal, quoted in ibid., p. 347.

p. 74   "On the west coast" A. Fuentes and B. Ehrenreich, quoted in ibid., p. 347.

p. 76   "we have combated from its incipiency . . ." in Elizabeth Faulkner Baker, *Technology and Women's Work,* p. 34.

cocoons

p. 77   "only in relation to the interminglings they make" Gilles Deleuze and Félix Guattari, *A Thousand Plateaus,* p. 90.

p. 78   "patient and monotonous efforts . . ." Fernand Braudel, *Capitalism and Material Life,* p. 244.

p. 78   "a collection of recipes . . ." ibid., p. 321.

p. 78   "the essence of femininity" Müntz, quoted in Sigmund Freud, "Leonardo da Vinci," *Art and Literature*, p. 201.

p. 79   "alien interest—in experimentation" ibid., p. 154.

p. 79   "work of the 'perspectors' . . ." Jean-Francois Lyotard, *The Postmodern Condition*, p. 44.

p. 79   "eccentric science . . ." Gilles Deleuze and Félix Guattari, *A Thousand Plateaus*, p. 361.

p. 80   "determined in such a way as to follow a flow" ibid., p. 409.

diagrams

p. 83   "It can imitate anything . . ." Karl Sigmund, *Games of Life*, p. 20.

p. 84   "something almost equally miraculous . . ." Andrew Hodges, *Alan Turing: The Enigma*, p. 109.

p. 85   "the mystery that woman represents . . ." Luce Irigaray, *Speculum of the Other Woman*, p. 26.

eve 1

p. 85   "uncovered female figures of silver" Charles Babbage, *Passages from the Life of a Philosopher*, p. 274.

p. 86   "Why not build a woman who should be . . ." Villiers de l'Isle Adam, *L'eve future*, p. 77.

p. 87   "electro-human creature" ibid., p. 103.

masterpieces

p. 88   "We like to believe" Alan Turing, quoted in Andrew Hodges, *Alan Turing: The Enigma*, p. 444.

p. 88   "the intention in constructing these machines" ibid., p. 356.

p. 88   "masters who are liable to get replaced" ibid., p. 357.

p. 89   "is to be regarded as nothing" ibid., pp. 377–78.

p. 89   "to copy the conscious mental processes" Hans Moravec, *Mind Children*, p. 16.

notes

trials

p. 90   "the part of B" was "taken by a man" Alan Turing, "On Computational Numbers," p. 422.

p. 90   "Answering questions with questions" Fah-Chun Cheong, *Internet Agents*, p. 278.

p. 91   "the apparent appropriateness and insight" Raymond Kurzweil, *The Age of Intelligent Machines*, p. 16.

p. 91   "USER: *'Men are all alike,'*" this and many other dialogues can be found—and conducted—on the Net.

p. 92   "is considered an improvement" Fah-Chun Cheong, *Internet Agents*, p. 253.

p. 92   "an agent more interesting than Eliza" ibid., p. 274.

p. 92   "adept at detecting and deflecting sexual advances" Sherry Turkle, *Life on the Screen*, p. 90.

p. 93   "it's not entirely clear to me" Leonard Foner, reference downloaded from the Net.

errors

p. 94   "If only you could see" *Blade Runner*, directed by Ridley Scott, 1982.

p. 95   "computing machines can only carry" Alan Turing, quoted in Andrew Hodges, *Alan Turing, The Enigma*, p. 358.

p. 95   "the more it schizophrenizes" Gilles Deleuze and Félix Guattari, *Anti-Oedipus*, p. 151.

eve 8

pp. 95–97   All quotations from *Eve of Destruction*, directed by Duncan Gibbons, 1991.

case study

p. 98   "The examination" Michel Foucault, *Discipline and Punish*, p. 191.

p. 98   "Discipline is a political anatomy of detail . . ." ibid., p. 139.

p. 98   "organized as a multiple, automatic and anonymous power . . ." ibid., p. 177.

p. 99   "an explosion of numerous . . ." Michel Foucault, *History of Sexuality, Volume 1*, p. 140.

p. 99   "lays down for each individual . . ." Michel Foucault, *Discipline and Punish*, p. 197.

p. 99   "meticulous observation of detail" Michel Foucault, *Discipline and Punish*, p. 141.

p. 99   "gradually learning what it meant . . ." Michel Foucault, *History of Sexuality, Volume 1*, p. 142.

what eve 8 next

p. 100   "the minute, I mean the nanosecond" William Gibson, *Neuromancer*, p. 132.

p. 100   "a reaction of this kind", Alan Turing, quoted in Andrew Hodges, *Alan Turing: The Enigma*, p. 357.

p. 101   "I am *both* bound" ibid., p. 473.

p. 101   "Went down to Sherborne" ibid., p. 484.

p. 102   "By the side of the bed" ibid., p. 488.

monster 1

p. 102   "Many and long were the conversations . . ." Mary Shelley, Preface to *Frankenstein*.

robotics

p. 103   "The signs on the office walls . . ." *The Economist*, Sept. 30, 1995, p. 107.

p. 104   "The problem is, of course, that it isn't a man" *The Economist,* May 18, 1996, p. 105.

learning curves

p. 104   "was every individual Man to divulge . . ." Mary Montagu, quoted in Dale Spender, *Women of Ideas and What Men Have Done to Them,* p. 76.

p. 105   Mary Astell, quoted in ibid., p. 63.

p. 105   "First as a mother" Comte, quoted in Michele le Doentt, *Philosophy and Psychoanalysis,* p. 190.

p. 105   "The transient trade we think evil" Charlotte Perkins Gilman, *Women and Economics,* p. 64.

p. 105   "Women—the deal. For—the game" Luce Irigaray, *Marine Lover,* p. 82.

p. 106   "And if you ask them insistently . . ." Luce Irigaray, *This Sex Which Is Not One,* p. 29.

p. 106   "Her mind is a matrix . . ." Misha, *"Wire Movement #9,"* p. 113.

p. 106   "A good woman does not have to be told . . ." Veronica Beechey and Elizabeth Whitelegg, eds. *Women in Britain Today,* p. 27.

p. 107   "Commodities, as we all know . . ." Luce Irigaray, *This Sex Which Is Not One,* p. 84.

p. 107   "if women are such good mimics" ibid., p. 76.

p. 107   *"what if these 'commodities' refused to go to 'market'?"* Luce Irigaray, *This Sex Which Is Not One,* p. 196.

p. 107   "They are all involved . . ." Jean Baudrillard, *Cool Memories,* p. 102.

p. 107   "Products are becoming digital" Donald Tapscott, *The Digital Economy,* p. 11.

p. 108   "It would be out of the question" Luce Irigaray, *This Sex Which Is Not One,* p. 110.

p. 108   "I learned fast" quoted in Cecilie Høigård and Liv Finstad, *Backstreets: Prostitution, Money and Love,* p. 83.

p. 109   " 'fluid' character" Luce Irigaray, *This Sex Which Is Not One,* p. 109.

anna o

p. 109   " 'hysterical' had become almost interchangeable" Elaine Showalter, *The Female Malady*, p. 129.

p. 109   "the devil . . ." Josef Breuer and Sigmund Freud, *Studies on Hysteria*, p. 332.

p. 110   "gaps in the memory" Sigmund Freud, *Case Histories 1, 'Dora' and 'Little Hans,'* pp. 46–47.

p. 110   "would complain of having 'lost' some time" Josef Breuer and Sigmund Freud, *Studies on Hysteria*, p. 76.

p. 110   "each of her momentary 'absences' " ibid., p. 318.

p. 110   "While everyone thought she was attending" ibid., p. 74.

p. 110   "Social circumstances often necessitate" ibid., p. 313.

p. 111   "Throughout the entire illness" ibid., p. 100.

p. 111   "the clearest intellect" ibid., p. 64.

p. 111   "an unusual degree of education and intelligence" ibid., p. 104.

p. 111   "bubbling over with intellectual vitality" ibid., p. 74.

p. 111   "an *excess* of efficiency" ibid., p. 313.

p. 111   "The overflowing productivity of their minds" ibid., p. 321.

p. 111   "double conscience" ibid., p. 64.

p. 111   "a whole number of activities" ibid., p. 313.

multiples

p. 113   "His responses had long since ceased to be a masquerade" Allucquére Rosanne Stone, *The War of Desire and Technology at the Close of the Mechanical Age*, p. 76.

switches

p. 114   "letters and telegrams are delivered with improbable despatch" Clive Leatherall, *Dracula*, p. 222.

p. 114   "Swan prepared some particularly fine thread . . ." W. A. Atherton, *From Compass to Computer: A History of Electrical and Electronics Engineering,* p. 132.

p. 115   "The news that the great experiment" Leonard de Vries, *Victorian Inventions,* pp 87–88.

speed queens

p. 117   "which adds, subtracts, multiplies . . ." in Elizabeth Faulkner Baker, *Technology and Woman's Work,* p. 213.

p. 117   "She adds the yards of the comptometer . . ." Heidi I. Hartmann et al., *Computer Chips and Paper Clips,* p. 73.

p. 117   "was approaching 2,000,000" Elizabeth Faulkner Baker, *Technology and Woman's Work,* p. 215.

p. 118   "I don't know about the world . . ." quoted in ibid., p. 71.

p. 118   "An English lady who demonstrated this machine" Leonard de Vries, *Victorian Inventions,* p. 166.

p. 119   "I have one in my office . . ." quoted in Heidi I. Hartmann et al., *Computer Chips and Paper Clips,* p. 26.

p. 119   "provided opportunity for a large number" ibid., p. 27.

p. 119   "earliest telephone companies" Bruce Sterling, *The Hacker Crackdown,* p. 12.

p. 120   "Basically, you, Miss Luthor" ibid., p. 29.

p. 120   "leads a very clear-cut" Gilles Deleuze and Félix Guattari, *A Thousand Plateaus,* p. 195.

p. 120   "the object of information" Michel Foucault, *Discipline and Punish,* p. 200.

p. 120   "The girl at the head of the line . . ." Elizabeth Faulkner Baker, *Technology and Woman's Work,* p. 215.

secrets

p. 122   "girl-less, cuss-less" Almon B. Strowger, quotes in W. A. Atherton, *From Compass to Computer,* p. 106.

p. 122   "contain moving parts that wear out . . ." Tom Duncan, *Electronics for Today and Tomorrow*, p. 195.

p. 123   "The specialized nature of their work . . ." Elizabeth Faulkner Baker, *Technology and Woman's Work*, p. 227.

p. 123   "a permanent inventiveness . . ." Gilles Deleuze and Félix Guattari, *A Thousand Plateaus*, p. 214.

p. 123   "In several exchanges reading clubs . . ." Elizabeth Faulkner Baker, *Technology and Woman's Work*, p. 70.

grass

p. 124   "A rhizome has no beginning" Gilles Deleuze and Félix Guattari, *A Thousand Plateaus*, p. 25.

p. 124   "Even when they have roots" ibid., p. 11.

p. 124   "Trees may correspond to a rhizome" ibid., p. 17.

p. 125   "no points or positions" ibid., p. 8.

p. 125   "Any point of a rhizome" ibid., p. 7.

p. 125   "may be broken" ibid., p. 9.

p. 125   "neither subject nor object" ibid., p. 8.

automata

p. 126   "A rich couple comes . . ." Gilles Deleuze and Félix Guattari, *A Thousand Plateaus*, p. 195.

p. 126   "extension of ear and voice . . ." Avital Ronell, *The Telephone Book*, p. 283.

p. 126   "invisible voices conducted . . ." ibid., pp. 301–2.

p. 126   "a fingertip mastery . . ." Elizabeth Faulkner Baker, *Technology and Woman's Work*, p. 242.

p. 126   "Having done this stuff a few hundred thousand times . . ." Bruce Sterling, *The Hacker Crackdown*, p. 30.

bugs

p. 127 "If computers are the power looms . . ." *Economist,* Oct. 29, 1994, p. 146.

p. 127 "There are a lot of cords down there . . ." Bruce Sterling, *The Hacker Crackdown,* p. 29.

p. 128 "working on behalf of their masters" *Economist, The World in 1995,* p. 143.

p. 128 "Computers can bring mathematical abstractions . . ." Hans Moravec, *Mind Children,* p. 133.

p. 128 "Agents are objects that don't wait to be told . . ." *Economist, The World in 1995,* p. 143.

p. 128 "with unknown consequences . . ." Fah-Chun Cheong, *Internet Agents,* p. 123.

p. 128 "of course all of this is quite abstract . . ." ibid., p. 122.

p. 129 "crawled through the network . . ." Richard B. Levin, *The Computer Virus Handbook,* p. 270.

p. 130 "spontaneously evolved, quite abstract . . ." Hans Moravec, *Mind Children,* p. 135.

disorders

p. 131 "My different personalities . . ." Anna Freud, quoted in Elisabeth Young-Bruehl, *Anna Freud,* p. 86.

p. 131 "battles and bargains . . ." ibid., p. 461.

p. 131 "Perhaps in the night I am a murderer" ibid., p. 58.

p. 132 "it irrupts in me, somehow . . ." ibid., p. 57.

p. 132 "lived as I did in the time . . ." ibid., p. 135.

p. 132 "no knowledge of each other or of the third . . ." Morton Prince, quoted in Roy Porter, ed. *The Faber Book of Madness,* p. 390.

p. 133 " 'The woman' who is Truddi Chase" Steven Shaviro, *Doom Patrols.* Available at http://dhalgren.english.washington.edu/~steve/doom.html

p. 134   "promoted by suggestion" Paul R. McHugh, "Multiple Personality Disorders." Available at http://www.psycom.het/mchugh

p. 134   "You are 'She,' I said . . ." Morton Prince, quoted in Roy Porter, ed. *The Faber Book of Madness*, p. 390.

p. 134   "if there is such a high degree of suggestive specificity . . ." Frank W. Putnam, debate with Paul R. McHugh. Available at http://www.asarian.org/nastrae/mpc_html/debate.html

p. 135   "a closed box, a unique entity, shut off from the others" Steven Shaviro, *Doom Patrols*. Available at http://dhalgren.english.washington.edu/~steve/doom.html

p. 135   "One of the things we hear from people . . ." Faith Christophe, "Can Selves Die?" Available at http://www.asarian.org/~astraea/household/

amazone

p. 137   "marriage-law lays it down" Herodotus, *The History of Herodotus*, Book IV, translated by George Rawlinson, Internet Classics Archive. Available at http://classics.mit.edu/Herodotus/history.html

p. 138   "The military art has no mystery" Mary Montagu, quoted in Dale Spender, *Women of Ideas*, p. 81.

p. 138   "The objective is not to gain ground" Monique Wittig, *Les Guérillères*, p. 95.

p. 138   "a Stateless woman-people" Gilles Deleuze and Félix Guattari, *A Thousand Plateaus*, p. 355.

p. 138   "fleshy passivity" Camille Paglia, *Sexual Personae*, pp. 75–77 passim.

p. 139   "come like fate, without reason, consideration, or pretext . . ." Gilles Deleuze and Félix Guattari, *A Thousand Plateaus*, p. 353.

p. 139   "could not tell what to make of the attack upon them" Herodotus, *The History of Herodotus*, Book IV. Available at http://classics.mit.edu/Herodotus/history.html

beginning again

p. 140   "Woman's desire . . ." Luce Irigaray, *This Sex Which Is Not One*, p. 25.

p. 140   "of *what materials* my *regiments* are to consist . . ." Ada Lovelace, October 1851, Betty A. Toole, *Ada, The Enchantress of Numbers*, p. 401.

p. 141   "Hysteria is silent and at the same time it mimes" Luce Irigaray, *This Sex Which Is Not One*, p. 137.

p. 141   " 'she' says something . . ." ibid., p. 103.

p. 141   "deep-going functional disorganization . . ." Josef Breuer and Sigmund Freud, *Studies on Hysteria*, pp. 78–79.

p. 142   "When they could thus understand one another" Herodotus, *The History of Herodotus*, Book IV. Available at http://classics.mit.edu/Herodotus/history.html

p. 143   "but 'within herself,' she never signs up" Luce Irigaray, *Marine Lover*, p. 90.

enigmas

p. 145   "A modern battleship" Marshall McLuhan, *Understanding Media*, p. 276.

p. 146   "close-up of the printer" reference downloaded from the Web.

p. 147   "the brains of Bletchley Park" F. H. Hinsley and Alan Stripp, eds. *Codebreakers: The Inside Story of Bletchley Park*, p. 65.

p. 147   "one of several 'men of the Professor type' " quoted in Andrew Hodges, *Alan Turing: The Enigma*, p. 206.

p. 147   "as far as making a pair of gloves" ibid., p. 207.

p. 147   "At this time" F. H. Hinsley and Alan Stripp, eds. *Codebreakers: The Inside Story of Bletchley Park*, p. 164.

p. 148   "greatly speeded up the routine solutions" ibid., p. 117.

p. 148   "Inevitably . . ." ibid., p. 115.

p. 148   "arrived at Bletchley Park . . ." ibid., p. 68.

p. 148   "life at sea, with the romantic idea . . ." ibid., p. 132.

p. 149   "twenty-two of us were drafted . . ." ibid., p. 133.

p. 149   "intricate complications of running the machines" ibid., p. 134.

p. 149   "a complicated drawing of numbers and letters" ibid., p. 136.

p. 150   "to put on to cards, with correct page references" ibid., p. 304.

p. 150   "I had buried this part of my life so completely" ibid., p. 137.

monster 2

p. 152   "I wanted to keep my software and use it over again" Grace Murray Hopper, quoted in Raymond Kurzweil, *The Age of Intelligent Machines*, p. 179.

spelling

p. 153   *"originate* anything" Ada Lovelace, notes to *Sketch of the Analytical Engine invented by Charles Babbage, Esq. By L. F. Menabrea, of Turin, Officer of the Military Engineers*, Note G.

p. 153   "curious, mysterious, marvellous, electrical, etc." Ada Lovelace, letter to Lady Byron, 1841, quoted in Dorothy Stein, *Ada, A Life and a Legacy*, p. 132.

p. 153   "unsensed forces" that "surround and influence us" Ada Lovelace, review of an "Abstract of 'Researches on Magnetism and on certain allied subjects,' including a supposed new Imponderable by Baron von Reichenbach," quoted in Dorothy Stein, *Ada, A Life and a Legacy*, p. 152.

p. 153   *"microscopical* structure" Ada Lovelace, August 1843, quoted in Betty A. Toole, *Ada, The Enchantress of Numbers*, p. 227.

p. 153   "to test certain points experimentally" Ada Lovelace, quoted in Dorothy Stein, *Ada, A Life and a Legacy*, p. 143.

p. 154   "I must be a most skillful *practical manipulator* . . ." Ada Lovelace, November 1844, quoted in Betty A. Toole, *Ada, The Enchantress of Numbers*, p. 295.

p. 154   "Could you ask the secretary . . ." Ada Lovelace, December 1844, quoted in Dorothy Stein, *Ada, A Life and a Legacy*, p. 149.

p. 154   "I am a *Fairy* you know" Ada Lovelace, January 1845, quoted in Betty A. Toole, *Ada, The Enchantress of Numbers*, p. 313.

hysteresis

p. 154   "Whether we examine distances travelled" Alvin Toffler, *Future Shock*, p. 33.

p. 154   "the advent of the telegraph" Marshall McLuhan, *Understanding Media*, p. 99.

p. 155   "Speed is the computer's secret weapon" T. R. Reid, *Microchip, the story of a revolution and the men who made it*, p. 21.

p. 155   "not at the target" Norbert Wiener, *Cybernetics*, p. 5.

p. 156   "Feedbacks of this general type" ibid., p. 113.

cybernetics

p. 156   "older machines, and in particular" Norbert Wiener, *The Human Use of Human Beings*, pp. 22–23.

p. 157   "effectively coupled to the external world" Norbert Wiener, *Cybernetics* p. 42.

p. 157   "keeps the engine from running wild . . ." Norbert Wiener, *The Human Use of Human Beings*, p. 152.

p. 157   "the first homeostatic machine in human history" Otto Mayr, *The Origins of Feedback*, p. 49.

p. 157   "the first nonliving object" Kevin Kelly, *Out of Control*, p. 113.

p. 158   "are precisely parallel" Norbert Wiener, *The Human Use of Human Beings*, p. 95.

p. 158   "the theory of the message" ibid., p. 27.

p. 158   "local and temporary islands" ibid., p. 36.

p. 158   "Life is an island here and now" ibid., p. 95.

p. 159   "It seems almost as if progress itself" ibid., pp. 46–47.

p. 159   "not yet spectators at the last stages of the world's death" ibid., p. 31.

p. 159   "Consider this principle of constancy" Luce Irigaray, *This Sex Which Is Not One*, p. 115.

p. 160   "involves sensory members" Norbert Wiener, *The Human Use of Human Beings*, p. 25.

p. 160   "No system is closed. The outside always seeps in . . ." Luce Irigaray, *This Sex Which Is Not One*, p. 116.

p. 160   "several possible sorts of behaviour" Gregory Bateson, *Mind and Nature*, p. 107.

p. 160   "positive gain, variously called *escalating* or *vicious* circles" ibid., p. 105.

p. 161   "Every intensity controls" Gilles Deleuze and Félix Guattari, *Anti-Oedipus*, p. 330.

p. 161   "the hint of death is present in every biological circuit" Gregory Bateson, *Mind and Nature*, p. 126.

p. 162   "If the open system is determined by anything" Antony Wilden, *System and Structure*, pp. 367–69.

p. 163   "feedback tends to oppose what the system is already doing" ibid., pp. 368–69.

p. 163   "once this exponential process" ibid., p. 367.

p. 163   "When the ecosystem is subjected" ibid., p. 75.

p. 163   "Any system-environment relationship" ibid., p. 369.

p. 164   "I always feel in a manner as if I *had* died" Ada Lovelace, March 1841, quoted in Doris Langley Moore, *Ada, Countess of Lovelace*, p. 98.

sea change

p. 165   "For a long time turbulence was identified with disorder or noise" Ilya Prigogine and Isabelle Stengers, *Order Out of Chaos*, p. 141.

p. 165   "How does a flow cross the boundary . . ." James Gleik, *Chaos*, p. 127.

p. 165   "The particles of cigarette smoke rise as one, for a while" ibid., p. 124.

p. 165   "fluctuations upon fluctuations, whorls upon whorls" ibid., p. 162.

p. 166   "A rope has been stretching" ibid., p. 127.

p. 166   "entities and variables" ibid., p. 108.

scattered brains

p. 166 "It does not appear to me that *cerebral* matter . . ." Ada Lovelace, November 1844, quoted in Betty A. Toole, *Ada, The Enchantress of Numbers*, p. 296.

p. 167 "Thought is not arborescent" Gilles Deleuze and Félix Guattari, *A Thousand Plateaus*, p. 15.

p. 168 "one concept will 'activate' another" Richard J. Eiser, *Attitudes, Chaos and the Connectionist Mind*, p. 30.

p. 168 " 'aha-experience' and the sudden 'insight' " Klaus Mainzer, *Thinking in Complexity*, p. 152.

p. 168 "Unlike a contact between two transistors" Christopher Wills, *The Runaway Brain*, p. 261.

p. 168 "When an axon of cell A" Donald Hebb, quoted in Klaus Mainzer, *Thinking in Complexity*, p. 126.

p. 169 "may be compared to an unnavigable river" Sigmund Freud, *Case Histories 1*, pp. 45–46.

p. 170 "in hysteria there is" Luce Irigaray, *This Sex Which Is Not One*, p. 138.

p. 170 "determinateness, direct logical analysis" Daniel McNeil and Paul Freiberger, *Fuzzy Logic*, p. 135.

p. 171 "overwork portions of their brains" *Economist*, Feb. 24, 1996, p. 124.

neurotics

p. 171 "at the end of the century" Alan Turing, quoted in Andrew Hodges, *Alan Turing, The Enigma*, p. 422.

p. 171 "the expert system market" Maureen Caudill and Charles Butler, *Naturally Intelligent Systems*, p. 26.

p. 172 "Victory seemed assured for the artificial sister" Raymond Kurzweil, *The Age of Intelligent Machines*, p. 149.

p. 173 "not only challenges" *Economist*, July 1, 1995, Internet survey.

p. 174 "achieved only limited success" Kevin Kelly, *Out of Control*, p. 296.

p. 174   "Parallel software is a tangled web" ibid., p. 308.

p. 174   "We are faced with a system" Richard J. Eiser, *Attitudes, Chaos and the Connectionist Mind*, p. 192.

p. 175   "arborescent and centralized" Gilles Deleuze and Félix Guattari, *A Thousand Plateaus*, p. 16.

p. 175   "transition machines" Andy Clark, *Associative Engines*, pp. 145–46.

intuition

p. 176   "I believe myself to possess almost singular combination of qualities . . ." Ada Lovelace, quoted in Betty A. Toole, *Ada, The Enchantress of Numbers*, p. 144.

p. 176   "On the human scale" T. R. Reid, *Microchip*, p. 21.

cave man

p. 177   "When men talk about virtual reality" Brenda Laurel, quoted in *Susie Bright's Sexual Reality: A Virtual Sex World Reader*, p. 67.

p. 178   "if we are ever to have pure knowledge" Plato, *The Last Days of Socrates*, pp. 110–12.

p. 178   "a metaphor of the inner space" Luce Irigaray, *Speculum of the Other Woman*, p. 243.

p. 179   "Illusion no longer has the freedom of the city" Luce Irigaray, *Marine Lover*, p. 99.

p. 179   "thus cuts himself off from the bedrock" Luce Irigaray, *Speculum of the Other Woman*, p. 133.

p. 179   "protection-projection screen" Luce Irigaray, *Marine Lover*, p. 87.

p. 179   "horror of nature is magicked away" ibid., p. 99.

p. 179   "you won't go wrong" Plato, *The Republic*, p. 517b.

p. 180   "a freedom that is limited only" Catherine Richards, "Virtual Bodies," p. 35.

p. 181   "you can lay Cleopatra" William Burroughs, *The Adding Machine*, p. 86.

p. 182   "the future will be a larger or greatly improved version of the *immediate past*," Marshall McLuhan, *The Gutenberg Galaxy*, p. 272.

p. 182   "Western man is externalizing himself in the form of gadgets" William Burroughs, *Naked Lunch*, p. 43.

hooked

p. 182   "All the forms of auxiliary apparatus" Sigmund Freud, "A Note upon the Mystic Writing Pad," *On Metapsychology*, p. 430.

p. 183   "living organism in its most simplified possible form" Sigmund Freud, "Beyond the Pleasure Principle," *On Metapsychology*, pp. 295–99.

p. 184   "And as they advance deeper out into the waves" Luce Irigaray, *Marine Lover*, p. 48.

tact

p. 185   "begins to be evident that 'touch' " Marshall McLuhan and Quentin Fiore, *War and Peace in the Global Village*, p. 71.

p. 185   "the extreme and pervasive tactility of the new electric environment" ibid., p. 77.

p. 185   "extraordinary technological clothing" Marshall McLuhan, *Understanding Media*, p. 159.

p. 185   "some kind of *actual space*" Larry McCaffrey, "Interview with William Gibson," quoted in Larry McCaffrey, ed. *Storming the Reality Studio*, p. 85.

p. 186   "the unclean promiscuity of everything" Jean Baudrillard, *The Ecstasy of Communication*, p. 27.

p. 186   "touch of the unknown" Elias Canetti, *Crowds and Power*, p. 15.

p. 186   "from a rhythmically pulsating environment" Lawrence K. Frank, "Tactile Communication," p. 202.

p. 186   "the weapon which lay nearest to hand" Elias Canetti, *Crowds and Power*, p. 248.

p. 186   "He wants to *see* what is reaching" ibid., p. 15.

p. 187   "One can picture the touch receptor" Ashley Montagu, *Touching: The Human Significance of the Skin*, p. 84.

p. 187   "Grooming the skin" Lawrence K. Frank, "Tactile Communication," p. 207.

p. 188   "possibility of distinguishing what is touching" Luce Irigaray, *This Sex Which Is Not One*, p. 26.

p. 188   "As in the case of taboo, the principal prohibition" Sigmund Freud, *The Origins of Religion*, p. 80.

p. 188   "When women talk about V.R." Brenda Laurel, quoted in *Susie Bright's Sexual Reality: A Virtual Sex World Reader*, p. 67.

p. 188   "the obstacle that separates thought from itself" Gilles Deleuze, *Cinema, Volume 2*, p. 189.

p. 188   "It is an entity so plugged in" Catherine Richards, "Virtual Bodies," p. 39.

p. 189   "elsewhere: another case of the persistence of 'matter' " Luce Irigaray, *This Sex Which Is Not One*, p. 125.

p. 189   "in what she says" ibid., p. 29.

p. 189   "tends to put the torch to fetish word" ibid., p. 79.

p. 189   "the human female" Ashley Montagu, *Touching: The Human Significance of the Skin*, p. 181.

p. 190   "have always turned . . ." Manon Regimbald, "The Borders of Textiles," p. 149.

p. 190   "Women have always spun" ibid., p. 147.

p. 190   "the computer as an electronic loom" Esther Parada, quoted in Trisha Ziff, "Taking new ideas back to the old world: talking to Esther Parada, Hector Méndez Caratini and Pedro Meyer," p. 132.

p. 190   "Until recently the computer" Cynthia Schira, "Powerful Creative Tools," p. 124.

p. 191   "are not only visual but also tactile" Louise Lemieux-Bérubé, "Textiles and New Technologies: a Common Language," p. 112.

cyberflesh

p. 192   "So I started to make a virtual body with a virtual wound" Linda Dement, "Screen Bodies," p. 9.

p. 193   "to conduct their dance" Michel Foucault, *Language, Counter-Memory, Practice*, p. 170.

mona lisa overdrive

p. 194   "the most perfect representation" Sigmund Freud, *Art and Literature*, p. 200.

p. 196   "hints that human activities once took place" Mary Rose Storey, *Mona Lisas*, p. 13.

p. 196   "From the start, he witnessed the harnessing of artistry to skilled engineering" Serge Bramly, *Leonardo, the Artist and the Man*, p. 71.

p. 196   "the application of many glazes" Mary Rose Storey, *Mona Lisas*, p. 14.

p. 197   "Her instincts of conquest" Sigmund Freud, "Leonardo da Vinci," *Art and Literature*, p. 201.

p. 197   "a sentence we may think his own" Serge Bramly, *Leonardo, the Artist and the Man*, p. 272.

p. 197   "often copying out word for word long passages" ibid., p. 270.

p. 197   "was much copied, *in toto* and in detail" ibid., p. 458n.

p. 198   "copying an existing machine" ibid., p. 271.

runaway

p. 199   "Her lover had asked her if she had come . . ." Linda Grant, *Sexing the Millennium*, p. 121.

p. 199   "Freud was right . . ." Baudrillard, *Seduction*, p. 6.

p. 199   "orgasms on one's own terms" ibid., p. 18.

p. 200   "Male orgasm had signified self-containment . . ." Donna Haraway, *Primate Visions*, p. 359.

p. 200   "universalistic claims made for human liberty . . ." Thomas Laquer, *Making Sex*, p. 150.

p. 200   "You will be organized" Gilles Deleuze and Félix Guattari, *A Thousand Plateaus*, p. 159.

p. 200   "most medical writers . . ." Donna Haraway, *Primate Visions*, p. 356.

p. 201   "ten years old, of delicate complexion . . ." Demetrius Zambaco, "Case History," p. 25.

p. 201   "cold showers . . ." ibid., p. 29.

p. 201   "Pubic belt, strait jacket . . ." ibid., p. 36.

p. 202   "To the analyst, any breakdown . . ." Roy Porter, *The Faber Book of Madness*, p. 488.

p. 203   "apologia for orgasm made by the Reichians . . ." Michel Foucault, quoted in David Macey, *The Lives of Michel Foucault*, p. 373.

p. 203   "I dismembered your body . . ." Alfonso Lingis, "Carnival in Rio," p. 61.

p. 203   "make of one's body . . ." Michel Foucault, quoted in James Miller, *The Passion of Michel Foucault*, p. 269.

p. 203   "the organic body, organized with survival as its goal" Jean-François Lyotard, *Libidinal Economy*, p. 2.

p. 203   "Flows of intensity . . ." Gilles Deleuze and Félix Guattari, *A Thousand Plateaus*, p. 162.

p. 204   "acentered systems . . ." Gilles Deleuze and Félix Guattari, *Anti-Oedipus*, p. 17.

p. 204   "we too are flows of matter and energy . . ." Manuel de Landa, "Nonorganic Life," p. 153.

p. 204   "Perforations occur in your body . . ." Monique Wittig, *The Lesbian Body*, p. 108.

p. 204   "Open the so-called body . . ." Jean-François Lyotard, *Libidinal Economy*, p. 2.

p. 204   "imply a multiplicity . . ." Gilles Deleuze and Félix Guattari, *A Thousand Plateaus*, p. 213.

p. 204   "inside every solitary living creature . . ." Kevin Kelly, *Out of Control*, p. 45.

p. 205   "As long as they do not threaten him" Elias Canetti, *Crowds and Power*, p. 420.

p. 205   "Do you know it is to me quite delightful . . ." Ada Lovelace, November 1844, quoted in Doris Langley Moore, *Ada, Countess of Lovelace*, p. 218.

p. 205   "nonhuman sex" Gilles Deleuze and Félix Guattari, *Anti-Oedipus*, p. 294.

p. 206   "does not take as its object" ibid., p. 292.

p. 206   "Not the clitoris or the vagina" Luce Irigaray, *This Sex Which Is Not One*, pp. 63–64.

p. 206   "a touching of *at least two* (lips)" ibid., p. 26.

p. 206   "the glans of the clitoris" Monique Wittig, *Les Guérillères*, p. 23.

p. 207   "making bits of bodies" Elizabeth Grosz, *Volatile Bodies*, p. 182.

p. 207   "question of 'passivity' is not the question of slavery" Jean-François Lyotard, *Libidinal Economy*, p. 260.

p. 207   "USE ME . . . what does she want" ibid., p. 66.

p. 207   "What interests the practitioners" Michel Foucault, "Sexual Choice, Sexual Act: Foucault and Homosexuality," p. 299.

p. 207   *"has sex organs . . ."* Luce Irigaray, *This Sex Which Is Not One*, p. 28.

p. 208   "exultation of a kind of autonomy . . ." Michel Foucault, quoted in James Miller, *The Passion of Michel Foucault*, p. 274.

p. 208   "Use me" Jean-François Lyotard, *Libidinal Economy*, p. 65.

p. 208   "sado-masochistic bond . . ." ibid., p. 63.

p. 208   "something 'unnameable,' 'useless' . . ." Michel Foucault, quoted in James Miller, *The Passion of Michel Foucault*, p. 274.

p. 208   "as though the watchman over our mental life . . ." Sigmund Freud, "The Economic Problem of Masochism," *On Metapsychology*, p. 413.

p. 208   "I stripped the will and the person . . ." Alfonso Lingis, "Carnival in Rio," p. 61.

p. 208   "the illusion of having no choice" Pat Califia, *Melting Point,* p. 172.

p. 208   "He wanted . . . everything" ibid., p. 108.

p. 209   "inventing new possibilities . . ." Michel Foucault, quoted in James Miller, *The Passion of Michel Foucault,* p. 263.

p. 209   "matter of a multiplication . . ." ibid., p. 274.

p. 209   "practices like fist-fucking . . ." ibid., p. 269.

p. 209   "becomes sheer ecstasy" ibid., p. 266.

p. 209   "Not even suffering . . ." Jean-François Lyotard, *Libidinal Economy,* p. 23.

p. 209   "That there are other ways . . ." Gilles Deleuze and Félix Guattari, *A Thousand Plateaus,* p. 55.

passing

p. 210   "Clothing himself in the female" Allucquére Rosanne Stone, "Invaginal Imaginal," p. 9.

p. 210   "To hide," write Deleuze and Guattari, "to camouflage oneself, is a warrior function" Gilles Deleuze and Félix Guattari, *A Thousand Plateaus,* p. 277.

p. 210   "To become the cyborg . . . is to put on the female" Allucquére Rosanne Stone, "Will the Real Body Please Stand Up?" p. 109.

p. 211   "Nowadays, it is said . . ." Avital Ronell, *The Telephone Book,* p. 302.

p. 212   "an intelligent machine would have to be intelligent enough . . ." Margaret A. Bowden, "Could a Robot Be Creative—and Would We Know?" quoted in Kenneth M. Ford, Clark Gylmour, and Patrick J. Hayes, eds. *Android Epistemology,* p. 256.

p. 212   "If we consider the great binary" Gilles Deleuze and Félix Guattari, *A Thousand Plateaus,* p. 213.

p. 213   "a microscopic transsexuality" ibid., p. 296.

p. 213   "Becoming-woman" ibid., p. 275.

p. 213   "overlooking the importance of imitation" ibid.

chemicals

p. 214   "tendency of biological research . . ." Sigmund Freud, "Leonardo da Vinci," in *Art and Literature*, p. 231.

p. 214   "characterized by its cyclic hormonal regulation . . ." Nelly Oudshoorn, *Beyond the Natural Body: an archeology of sex hormones*, p. 146.

p. 215   "The causes are not yet defined . . ." *Birmingham Post*, Mar. 8, 1996, p. 10.

p. 216   "didn't kill the roosters . . ." Theo Colborn, Dianne Dumanoski, and John Peterson Myers, *Our Stolen Future*, p. 199.

p. 216   "plastics are not inert . . ." ibid., pp. 233–34.

p. 216   "These ubiquitous metal cans attached to electrical poles . . ." ibid., p. 92.

p. 216   "a striking inverse correlation . . ." ibid., p. 175.

p. 217   "consorts so readily with foreigners . . ." ibid., p. 70.

p. 217   "The body responds to the impostors . . ." ibid., p. 205.

p. 217   "impersonate them . . ." ibid., p. 204.

p. 217   "chemicals interfering with hormonal messages . . ." ibid., p. 195.

p. 217   "Without these testosterone signals . . ." ibid., p. 83.

p. 218   "not only on what the mother takes . . ." ibid., p. 212.

p. 218   "By disrupting hormones and development . . ." ibid., p. 197.

p. 218   "undermine the ways in which humans interact . . ." ibid., p. 207.

p. 218   "alter the characteristics that make us uniquely human . . ." ibid., p. 234.

p. 218   "the legacy of our species . . ." ibid., p. 238.

xyz

p. 218   "Sex is not a necessary condition for life . . ." François Jacob, *The Logic of Life*, p. 356.

p. 219   "frozen accident" Steven Levy, *Artificial Life: The Quest for a New Creation*, p. 198.

p. 220   "it would be a small matter for dandelions to sprout butterfly wings" Dorion Sagan, "Metametazoa," p. 378.

the peahen's tale

p. 223   "preservation of favourable variations and the rejection of injurious variations" Charles Darwin, quoted in François Jacob, *The Logic of Life*, p. 170.

p. 223   "It may be said that natural selection" Charles Darwin, quoted in ibid., p. 171.

p. 223   "is a game with its own rules" ibid., p. 296.

p. 224   "When the males and females of any animal have the same general habits of life" ibid., p. 172.

p. 225   "virility tests designed to get most males" *Economist*, Dec. 23, 1995–Jan. 5, 1996, p. 121.

p. 226   "is not profited personally by his mane" Charlotte Perkins Gilman, *Women and Economics*, p. 33.

p. 226   "resembles artificial breeding" Karl Sigmund, *Games of Life*, p. 126.

p. 227   "the further development of the plumage character" R. A. Fisher, *The Genetical Theory of Natural Selection*, p. 152.

p. 227   "rides, like a surfer, on a wave of ever-increasing tail lengths sweeping through the population" Karl Sigmund, *Games of Life*, p. 131.

p. 227   "The two characteristics affected by such a process" R. A. Fisher, *The Genetical Theory of Natural Selection*, p. 152.

p. 228   "Where one function is carried to unnatural excess" Charlotte Perkins Gilman, *Women and Economics*, p. 72.

p. 228   "All morbid conditions tend to extinction" ibid.

p. 228   "So it was female choice which caused the males' long tails" Karl Sigmund, *Games of Life*, p. 126.

p. 228   "in many species, females had a large say" Matt Ridley, *The Red Queen*, p. 133.

loops

p. 230   "the egg uses the messages passed on" Stephen S. Hall, *Mapping the Next Millennium*, p. 212.

p. 230   "Eggs are computers to the simple floppy discs of sperm" *Economist*, Sept. 3, 1994, p. 91.

p. 230   "are not organizers, but mere inductors" Gilles Deleuze and Félix Guattari, *Anti-Oedipus*, p. 91.

p. 230   "Doubtless one can *believe* that, in the beginning (?)" ibid., p. 92.

p. 231   "these machines have to be built first. For this, a blueprint is required" Karl Sigmund, *Games of Life*, p. 73.

p. 232   "numerous and daring experiments in parthenogenesis" Simone de Beauvoir, *The Second Sex*, p. 34.

p. 232   "The geneticists first realized that F.D. was unusual" *New Scientist*, Oct. 7, 1995, p. 17.

symbionts

p. 234   "the greatest pollution crisis the earth has ever known" Dorion Sagan, "Metametazoa: Biology and Multiplicity," p. 367.

p. 234   "each eukaryotic 'animal' cell . . ." Dorion Sagan, "Metametazoa," p. 363.

p. 235   "biochemically and metabolically far more diverse . . ." Dorion Sagan, "Metametazoa," p. 377.

p. 235   "as people have passed through since they were apes," Matt Ridley, *The Red Queen*, p. 64.

p. 235   "four fifths of the history of life on Earth has been solely a bacterial phenomenon" Dorion Sagan, "Metametazoa," p. 377.

p. 235   "most salient feature . . ." Stephen Jay Gould, "The Evolution of Life on the Earth," p. 65.

p. 236   "blend into a pointillist landscape" Dorion Sagan, "Metametazoa," p. 363.

p. 236   "With bacteria, unlike organisms . . ." François Jacob, *The Logic of Life*, p. 297.

eve 2

p. 237   "As a result, the inheritance of mitochondrial chromosomes" Christopher Wills, *Runaway Brain*, p. 23.

p. 238   "the woman who is the most recent direct ancestor" Daniel Dennett, *Darwin's Dangerous Idea*, p. 97.

pottering

p. 239   "a well-established dictum . . ." Margaret Alic, *Hypatia's Heritage*, p. 111.

p. 239   "ideas, taste, and elegance" Hegel, quoted in Michèle Le Doeuff, "Philosophy and Psychoanalysis," pp. 189–90.

p. 240   "is suddenly full of boulders and screes and speculations about strata" Margaret Lane, *The Tale of Beatrix Potter*, p. 41.

p. 240   "the precise and the minute" ibid., p. 40.

p. 241   "composed of fungal cells" Mark McMenamin and Dianna McMenamin, *Hypersea*, p. 68.

p. 241   "People take symbiosis seriously in lichens" Lynn Margulis, quoted in ibid., p. 67.

p. 241   "complex overgrown lichens" ibid., p. 171.

p. 241   "so genetically open" Dorion Sagan, "Metametazoa," p. 378.

p. 242   "fluid genetic transfers" ibid., p. 363.

p. 242   "without identifiable terms, without accounts" Luce Irigaray, *This Sex Which Is Not One*, p. 197.

p. 242   "The body can no longer . . ." Dorion Sagan, "Metametazoa," p. 368.

p. 242   "With bacteria . . ." François Jacob, *The Logic of Life*, p. 297.

p. 242   "To be woman, she does not have to be mother," Luce Irigaray, *Marine Lover*, p. 86.

mutants

p. 244   "Joan explained how she had been taught" Andrew Hodges, *Alan Turing, The Enigma,* p. 208.

p. 244   "always enjoyed examining plants" ibid., p. 434.

p. 245   "a slender but subversive strand" Stephen S. Hall, *Mapping the Next Millennium,* p. 216.

p. 246   "the enzyme cuts one strand of the DNA helix" ibid., p. 232.

p. 246   "leftover from the merging of stranger bacteria" Dorion Sagan, "Metametazoa," p. 378.

p. 246   "quasispecies," "swarms," or "consensus sequences" Laurie Garrett, *The Coming Plague,* p. 579.

p. 246   "individual alteration can change an entire system . . ." ibid., p. 619.

p. 247   "the ability to outwit or manipulate" ibid., p. 618.

p. 247   "We form a rhizome with our viruses" Gilles Deleuze and Félix Guattari, *A Thousand Plateaus,* p. 10.

wetware

p. 248   "think of the sea from afar" Luce Irigaray, *Marine Lover,* p. 51.

p. 248   "have far more to them . . ." James Lovelock, *Gaia,* p. 78.

p. 248   "on the land is for the most part two-dimensional . . ." ibid., p. 87.

p. 248   "terrestrial organisms had to build for themselves . . ." Mark McMenamin and Dianna McMenamin, *Hypersea,* p. 4.

p. 248   "biota has had to find ways to carry the sea within it" ibid., p. 5.

p. 249   "Acting over evolutionary time as a rising tide . . ." p. 25.

p. 249   "terrestrial sea . . ." ibid., p. 93.

p. 249   "the appearance of complex life on land was a major event" ibid., p. 25.

p. 249   "Trees are neither found nor needed in the sea . . ." James Lovelock, *Gaia,* p. 87.

p. 249   "numerically dominated by tiny single-celled *protista* . . ." ibid., p. 88.

p. 249   "from the first appearance of marine bacteria . . ." Mark McMenamin and Dianna McMenamin, *Hypersea*, p. 73.

p. 250   "highly flattened fronds, sheets and circlets" Stephen Jay Gould, *Scientific American*, October 1994, p. 67.

dryware

p. 250   "an island, enclosed by nature itself within unalterable limits" Immanuel Kant, *Critique of Pure Reason*, p. 257.

p. 250   "If a man wants to delude himself" Luce Irigaray, *Marine Lover*, p. 46.

p. 250   "delivers man to the uncertainty of fate" Michel Foucault, *Madness and Civilization*, pp. 10–12.

p. 250   "One thing at least is certain" ibid., p. 18.

p. 250   "the river with its thousand arms" pp. 10–12.

p. 251   "If only the sea did not exist" Luce Irigaray, *Marine Lover*, p. 51.

quanta

p. 252   *"made* capacity where there technically wasn't any" Pat Cadigan, *Synners*, p. 174.

p. 253   "midway between the fluid and the solid" Elias Canetti, *Crowds and Power*, p. 101.

p. 253   *"age of sand"* Donald Tapscott, *Digital Economy, Promise and Peril in the Age of Networked Intelligence*, p. 48.

casting off

p. 256   "little bit of another sense . . ." Ada Lovelace, March 1841, quoted in Doris Langley Moore, *Ada, Countess of Lovelace*, p. 98.

p. 256   "further extension" Ada Lovelace, September 1841, quoted in Dorothy
          Stein, *Ada, A Life and a Legacy*, p. 79.

p. 256   "Perhaps none of us can estimate . . ." Ada Lovelace, February 1840,
          quoted in Doris Langley Moore, *Ada, Countess of Lovelace*, p. 96.

p. 256   "thunderstruck by the power of the writing . . ." Ada Lovelace, July
          1843, quoted in Dorothy Stein, *Ada, A Life and a Legacy*, p. 110.

bibliography

A.A.L., "Notes to Sketch of the Analytical Engine invented by Charles Babbage Esq. By L. F. Menabrea, of Turin, Officer of the Military Engineers," in Morrison, Philip and Emily, eds. *Charles Babbage and his Calculating Engines: Selected Writings by Charles Babbage and others.* New York: Dover, 1961.

Acker, Kathy. *Empire of the Senseless.* New York: Grove Press, 1988.

Alic, Margaret. *Hypatia's Heritage.* London: The Women's Press, 1990.

Atherton, W. A. *From Compass to Computer, A History of Electrical and Electronics Engineering.* London: Macmillan, 1984.

Atwood, Margaret. *The Handmaid's Tale.* London: Virago Press, 1995.

Babbage, Charles. *Passages from the Life of a Philosopher.* London: William Pickering, 1994.

Barber, Elizabeth Wayland. *Women's Work; The First 20,000 Years.* New York: W. W. Norton & Co., 1994.

Bateson, Gregory. *Mind and Nature. A Necessary Unity.* New York: Dutton, 1979.

Baudrillard, Jean. *Cool Memories.* London: Verso, 1990.

———. *Seduction.* London: Macmillan, 1990.

———. *The Ecstasy of Communication.* New York: Semiotext(e), 1988.

Beechey, Veronica, and Elizabeth Whitelegg, eds. *Women in Britain Today.* Philadelphia: Open University Press, 1986.

Bowden, Margaret A. "Could a Robot Be Creative—and Would We Know?" in Kenneth M. Ford, Clark Gylmour, and Patrick J. Hayes, eds. *Android Epistemol-*

ogy. Menlo Park and Cambridge, Mass.: American Association for Artificial Intelligence and MIT, 1995.

Bramly, Serge. *Leonardo, The Artist and the Man*. London: Penguin, 1994.

Braudel, Fernand. *Capitalism and Material Life 1400–1800*. London: Weidenfeld and Nicolson, 1973.

Breuer, Josef, and Sigmund Freud. *Studies on Hysteria*. London: Penguin, 1991.

Briggs, Asa. *The Age of Improvement*. London: Longmans, 1959.

Bright, Susie. *Susie Bright's Sexual Reality: A Virtual Sex World Reader*. Pittsburgh: Cleis Press, 1992.

Burroughs, William. *The Adding Machine: Collected Essays*. London: John Calder, 1985.

————. *Naked Lunch*. London: Corgi Books, 1968.

Butler, Octavia. *Dawn. Xenogenesis:l*. London: Victor Gollancz, 1988.

Cadigan, Pat. *Synners*. London: HarperCollins, 1991.

————. *Fools*. London: HarperCollins, 1994.

Califia, Pat. *Melting Point*. Boston: Alyson Publications, 1993.

Canetti, Elias. *Crowds and Power*. London: Penguin, 1984.

Caudill, Maureen, and Charles Butler. *Naturally Intelligent Systems*. Cambridge, Mass.: MIT, 1991.

Cheong, Fah-Chun. *Internet Agents: Spiders, Wanderers, Brokers, and Bots*. Indianapolis: New Riders, 1996.

Clark, Andy. *Associative Engines. Connectionism, Concepts, and Representational Change*. Cambridge, Mass.: MIT, 1993.

Cocteau, Jean. *Opium*. London: Peter Owen, 1990.

Colborn, Theo, Dianne Dumanoski, and John Peterson Myers. *Our Stolen Future*. London: Little, Brown & Co., 1996.

Daly, Mary. *Gyn/Ecology, The Metaethics of Radical Feminism*. London: The Women's Press, 1979.

de Beauvoir, Simone. *The Second Sex*. London: Jonathan Cape, 1960.

de Landa, Manuel. "Non-organic Life," in Crary, Jonathan and Sanford Kwinter, eds. *Incorporations, Zone 6*. New York: Zone Books, 1992.

————. *War in the Age of Intelligent Machines*. New York: Zone Books, 1991.

Debord, Guy. *Comments on the Society of the Spectacle*. London: Verso, 1988.

Deleuze, Gilles. *Cinema, 2, The Time-Image*. London: Athlone Press, 1989.

————. *The Fold, Leibniz and the Baroque*. London: Athlone Press, 1993.

————. *Difference and Repetition*. London: Athlone Press, 1994.

Deleuze, Gilles and Félix Guattari. *A Thousand Plateaus: Capitalism and Schizophrenia*. London: Athlone, 1988.

————. *Anti-Oedipus: Capitalism and Schizophrenia*. London: Athlone, 1990.

Dement, Linda. "Screen Bodies," *Women's Art,* No. 63, March/April 1995.

———. *Cyberflesh Girlmonster* (CD-ROM). 1995.

Dennett, Daniel. *Darwin's Dangerous Idea: Evolution and the Meanings of Life.* London: Penguin, 1995.

Dicken, Peter. *Global Shift, The Internationalization of Economic Activity.* London: Paul Chapman, 1992.

Duncan, Tom. *Electronics for Today and Tomorrow.* London: John Murray, 1993.

Eiser, J. Richard. *Attitudes, Chaos and the Connectionist Mind.* Oxford: Blackwell, 1994.

Eliade, Mircea. *Rites and Symbols of Initiation: The Mysteries of Birth and Rebirth.* New York: Harper Torchbooks, 1965.

English, W. *The Textile Industry.* London: Longmans, 1969.

Faulkner Baker, Elizabeth. *Technology and Woman's Work.* New York: Columbia University Press, 1966.

Fisher, R. A. *The Genetical Theory of Natural Selection.* New York: Dover, 1958.

Foucault, Michel. *Discipline & Punish. The Birth of the Prison.* New York: Vintage Books, 1995.

———. *The Archeology of Knowledge.* London: Tavistock Publications, 1978.

———. *Language, Counter-Memory, Practice.* Ithaca, New York: Cornell University Press, 1977.

———. *Madness and Civilization, A History of Insanity in the Age of Reason.* New York: Pantheon, 1965.

———. *History of Sexuality, Volume 1: An Introduction.* New York: Pantheon, 1978.

———. "Sexual Choice, Sexual Act: Foucault and Homosexuality," in Lawrence D. Kritzman, ed. *Politics, Philosophy, Culture, Interviews and Other Writings, 1977–1984.* London: Routledge, 1988.

Frank, Lawrence K. "Tactile Communication," in Smith, Alfred E., ed. *Communication and Culture.* New York: Holt, Rinehart and Winston, 1966.

Frazer, J. G. *The Golden Bough. A Study in Magic and Religion.* London: Macmillan, 1974.

Freud, Sigmund. *New Introductory Lectures on Psychoanalysis.* Penguin Freud Library Volume 2. London: Penguin, 1977.

———. *On Sexuality.* Penguin Freud Library Volume 7. London: Penguin, 1977.

———. *On Metapsychology: The Theory of Psychoanalysis.* Penguin Freud Library Volume 11. London: Penguin, 1984.

———. *Case Histories I.* Penguin Freud Library Volume 8. London: Penguin, 1977.

———. *The Origins of Religion.* Penguin Freud Library Volume 13, London: Penguin, 1990.

————. *Art and Literature*. Penguin Freud Library Volume 14. London: Penguin, 1990.

Garrett, Laurie. *The Coming Plague; Newly Emerging Diseases in a World Out of Balance*. London: Penguin, 1995.

Gibson, William. *Neuromancer*. New York: Ace Science Fiction, 1984.

————. *Count Zero*. London: Grafton, 1987.

————. *Mona Lisa Overdrive*. London: Grafton, 1989.

Gibson, William, and Bruce Sterling. *The Difference Engine*. London: Victor Gollancz, 1992.

Ginzberg, Carlo. *Ecstasies. Deciphering the Witches' Sabbath*. London: Hutchinson Radius, 1990.

Gleick, James. *Chaos, Making a New Science*. London: Sphere Books, 1991.

Gould, Stephen Jay. "The Evolution of Life on the Earth," *Scientific American*, October 1994, Volume 27, No. 4, pp. 53–61.

Grant, Linda. *Sexing the Millennium: A Political History of the Sexual Revolution*. London: HarperCollins, 1994.

Grosz, Elizabeth. *Volatile Bodies. Toward a Corporeal Feminism*. Bloomington: Indiana University Press, 1994.

Hall, Stephen S. *Mapping the Next Millennium*. New York: Vintage Books, 1993.

Haraway, Donna J. "A Cyborg Manifesto: Science, Technology, and Socialist Feminism in the Late Twentieth Century," in *Simians, Cyborgs, and Women, the Reinvention of Nature*. London: Free Association Books, 1991.

————. *Primate Visions*. London: Verso, 1992.

Hartman Strom, Sharon. "Machines Instead of Clerks: Technology and the Feminization of Bookkeeping 1910–1950," in Hartmann, Heidi I. *Computer Chips and Paper Clips, Technology and Women's Employment, Volume II*. Washington: National Academic Press, 1987.

Hartmann, Heidi I., Robert E. Kraut, and Louise A. Tilly, eds. *Computer Chips and Paper Clips, Technology and Women's Employment, Volume I*. Washington: National Academic Press, 1986.

Herodotus. *The Histories*. Harmondsworth: Penguin, 1996. Also available as *The History of Herodotus*, Book IV, in the Internet Classics Archive at http://classics.mit.edu/Herodotus/history.html.

Hinsley, F. H., and Alan Stripp, eds. *Codebreakers: The Inside Story of Bletchley Park*. Oxford: Oxford University Press, 1994.

Hodges, Andrew. *Alan Turing: The Enigma*. New York: Simon and Schuster, 1983.

Høigård, Cecilie, and Liv Finstad. *Backstreets: Prostitution, Money and Love*. Cambridge: Polity Press, 1992.

Hollingdale, S. H., and G. C. Tootill. *Electronic Computers*. London: Penguin, 1982.

Irigaray, Luce. *Marine Lover of Friedrich Nietzsche*. New York: Columbia University Press, 1991.

———. *Speculum of the Other Woman*. New York: Cornell University Press, 1992.

———. *This Sex Which Is Not One*. New York: Cornell University Press, 1993.

Jacob, François. *The Logic of Life: A History of Heredity, and The Possible and the Actual*. London: Penguin, 1982.

James, Carol L., and Duncan E. Morrill. "The Real Ada; Countess of Lovelace," *ACM SIGsoft Software Engineering Notes*, Vol. 8, No. 1, Jan. 1983.

Jennings, Humphrey. *Pandaemonium: The Coming of the Machine as Seen by Contemporary Observers*. London: Andre Deutsch, 1985.

Kant, Immanuel. *The Critique of Pure Reason*. London: MacMillan, 1990.

Kelly, Kevin. *Out of Control: The New Biology of Machines*. London: Fourth Estate, 1994.

Klingender, Francis D. *Art and the Industrial Revolution*. St. Albans: Paladin, 1975.

Kramer, Henrich and James Sprenger. *Malleus Maleficarum*. London: Arrow Books, 1971.

Kurzweil, Raymond. *The Age of Intelligent Machines*. Cambridge, Mass.: MIT, 1990.

Lacan, Jacques. *The Works of Jacques Lacan*. London: Free Association Books, 1986.

Landow, George P. *Hypertext: The Convergence of Contemporary Critical Theory and Technology*. Baltimore: Johns Hopkins University Press, 1992.

Lane, Margaret. *The Tale of Beatrix Potter*. London: Penguin, 1985.

Laqueur, Thomas. *Making Sex: Body and Gender from the Greeks to Freud*. Cambridge, Mass.: Harvard University Press, 1992.

Leatherall, Clive. *Dracula: The Novel and the Legend*. Wellingborough: Acquarian, 1985.

Le Corbusier. *The City of Tomorrow and Its Planning*. London: Architectural Press, 1947.

Le Doeuff, Michèle. "Philosophy and Psychoanalysis," in Moi, Toril, ed. *French Feminist Thought*. Oxford: Blackwell, 1987.

Lemieux-Bérubé, Louise. "Textiles and New Technologies: A Common Language," in *Textiles Sismographes*. Texts from the *Symposium Fibres et Textiles*. Montréal: Conseil des Arts Textiles du Quebec, 1995.

Levin, Richard B. *The Computer Virus Handbook*. London: Osborne McGraw-Hill, 1990.

Levy, Steven. *Artificial Life: The Quest for a New Creation*. London: Penguin, 1993.

Lingis, Alfonso. "Carnival in Rio," in *Vulvamorphia, Lusitania #6*. New York, 1994.

Lovelock, James. *Gaia, A New Look at Life on Earth*. Oxford: Oxford University Press, 1995.

Lupton, Ellen. *Mechanical Brides: Women and Machines from Home to Office*. New York: Cooper-Hewitt and Princeton Architectural Press, 1993.

Lyotard, Jean-François. *Libidinal Economy*. London: Athlone Press, 1993.

———. *The Postmodern Condition: A Report on Knowledge*. Manchester: Manchester University Press, 1984.

Macey, David. *The Lives of Michel Foucault*. London: Vintage, 1994.

Mainzer, Klaus. *Thinking in Complexity. The Complex Dynamics of Matter, Mind, and Mankind*. Berlin: Springer-Verlag, 1994.

Mayr, Otto. *The Origins of Feedback Control*. Cambridge, Mass.: MIT, 1968.

McCaffery, Larry. "An Interview with William Gibson," in McCaffery, Larry, ed. *Storming the Reality Studio*. London: Duke University Press, 1991.

McLuhan, Marshall. *Understanding Media. The Extensions of Man*. London: Sphere Books, 1969.

———. *The Gutenberg Galaxy: The Making of Typographic Man*. London: Routledge and Kegan Paul, 1962.

———. and Quentin Fiore. *War and Peace in the Global Village*. New York: Bantam Books, 1967.

McMenamin, Mark, and Dianna McMenamin. *Hypersea: Life on Land*. New York: Columbia University Press, 1994.

McNeil, Daniel, and Paul Freiberger. *Fuzzy Logic*. New York: Touchstone, 1994.

Miller, James. *The Passion of Michel Foucault*. London: HarperCollins, 1993.

Misha. "Wire Movement #9," in McCaffery, Larry, ed. *Storming the Reality Studio*. London: Duke University Press, 1991.

Mitchell, Juliet, and Jacqueline Rose, eds. *Feminine Sexuality, Jacques Lacan and the Ecole Freudienne*. London: Macmillan, 1982.

Montagu, Ashley. *Touching: The Human Significance of the Skin*. New York: Columbia University Press, 1971.

Moore, Doris Langley. *Ada, Countess of Lovelace, Byron's Illegitimate Daughter*. London: John Murray, 1977.

Moravec, Hans. *Mind Children: The Future of Robot and Human Intelligence*. Cambridge, Mass.: Harvard University Press, 1988.

Morrison, Philip and Emily, eds. *Charles Babbage and his Calculating Engines: Selected Writings by Charles Babbage and Others*. New York: Dover, 1961.

Naisbitt, John. *Megatrends Asia*. London: Nicholas Brealey, 1996.

Oudshoorn, Nelly. *Beyond the Natural Body: an archeology of sex hormones*. London: Routledge, 1994.

Paglia, Camille. *Sexual Personae: Art and Decadence from Nefertiti to Emily Dickinson*. London: Penguin, 1991.

Perkins Gilman, Charlotte. *Herland*. London: The Women's Press, 1979.

———. *Women and Economics: A Study of the Economic Relation between Men and Women as a Factor in Social Evolution*. New York: Harper Torchbooks, 1966.

Plato. *The Last Days of Socrates*. London: Penguin, 1975.

———. *The Republic*. London: Penguin, 1975.

Porter, Roy, ed. *The Faber Book of Madness*. London: Faber and Faber, 1993.

Prigogine, Ilya, and Isabelle Stengers. *Order Out of Chaos*. Flamingo: London, 1985.

Regimbald, Manon. "The Borders of Textiles," in *Textiles Sismographes*. Texts from the *Symposium Fibres et Textiles*. Montréal: Conseil des Arts Textiles du Quebec, 1995.

Reid, T. R. *Microchip, The story of a revolution and the men who made it*. London: Collins, 1985.

Richards, Catherine. "Virtual Bodies," in de Guerre, Marc, and Kathleen Pirrie Adams, eds. *Throughput, Public 11*. Toronto: Public Access, 1995.

Ridley, Matt. *The Red Queen. Sex and the Evolution of Human Nature*. London: Penguin, 1994.

Ronell, Avital. *The Telephone Book; Technology, Schizophrenia, Electric Speech*. London: University of Nebraska Press, 1989.

Sagan, Dorion. *Biospheres*. London: Arkana, 1980.

———. "Metametazoa," in Crary, Jonathan and Sanford Kwinter, eds. *Incorporations, Zone 6*. New York: Zone Books, 1992.

Schaffer, Simon. "Babbage's Dancer and the Impressarios of Mechanism," in Spufford, Francis and Jenny Uglow, eds. *Cultural Babbage. Technology, Time and Invention*. London: Faber and Faber, 1996.

Schira, Cynthia. "Powerful Creative Tools," in *Textiles Sismographes*. Texts from the *Symposium Fibres et Textiles*. Montréal: Conseil des Arts Textiles du Quebec, 1995.

Shaviro, Steven. *Doom Patrol*. London: Serpent's Tail, 1996.

Shelley, Mary. *Frankenstein*. London: Penguin, 1992.

Showalter, Elaine. *The Female Malady, Women, Madness and English Culture, 1830–1980*. London: Virago, 1995.

Sigmund, Karl. *Games of Life. Explorations in Ecology, Evolution, and Behaviour*. Oxford: Oxford University Press, 1993.

Smith, Stevie. *The Holiday*. London: Virago, 1979.

Spender, Dale. *Women of Ideas and What Men Have Done to Them*. London: Ark, 1983.

Stein, Dorothy. *Ada, A Life and a Legacy*. Cambridge, Mass.: MIT, 1985.

Sterling, Bruce. *The Hacker Crackdown, Law and Order on the Electronic Frontier*. London: Penguin, 1994.

Stone, Allucquére Rosanne. "Will the Real Body Please Stand Up?," in Benedikt, Michael, ed. *Cyberspace, First Steps*. Cambridge, Mass.: MIT, 1994.

———. *The War of Desire and Technology at the Close of the Mechanical Age*. Cambridge, Mass.: MIT, 1995.

———. "Invaginal Imaginal," in *Vulvamorphia, Lusitania #6*. New York, 1994.

Storey, Mary Rose. *Mona Lisas*. London: Constable, 1980.

Sun-Tzu, *The Art of War*. New York: William Morrow, 1993.

Tapscott, Donald. *The Digital Economy, Promise and Peril in the Age of Networked Intelligence*. New York: McGraw-Hill, 1996.

Toffler, Alvin. *Future Shock*. London: Pan Books, 1971.

Toole, Betty A. *Ada, the Enchantress of Numbers*. California: Strawberry Press, 1992.

Turing, Alan. "On Computational Numbers." *Mind: A Quarterly Review of Psychology and Philosophy*, October 1950. Volume LIX, No. 236.

Turkle, Sherry. *Life on the Screen, Identity in the Age of the Internet*. London: Weidenfeld and Nicolson, 1996.

Villiers de l'Isle Adam. *L'Eve Future*. Paris: Jean-Jacques Pauvert, 1960.

de Vries, Leonard. *Victorian Inventions*. London: John Murray, 1973.

Wiener, Norbert. *Cybernetics: Communication and Control in Animal and Machine*. Cambridge, Mass.: MIT, 1948.

———. *The Human Use of Human Beings. Cybernetics and Society*. London: Eyre and Spottiswood, 1954.

Wilden, Antony. *System and Structure*. London: Tavistock, 1972.

Wilkinson, Helen. *No Turning Back: Generations and the Genderquake*. London: Demos, 1994.

Wills, Christopher. *The Runaway Brain: The Evolution of Human Uniqueness*. London: Flamingo, 1995.

Wittig, Monique. *Les Guérillères*. Boston: Beacon Press, 1985.

———. *The Lesbian Body*. Boston: Beacon Press, 1986.

———. *The Straight Mind*. Hemel Hempstead: Harvester Wheatsheaf, 1992.

Woolf, Virginia. *Orlando*. London: Penguin, 1993.

Young-Bruehl, Elisabeth. *Anna Freud*. London: Papermac, 1992.

Zambaco, Demetrius. "Case History," *Polysexuality. Semiotext(e) 10,* New York, 1981.

Ziff, Trisha. "Taking new ideas back to the old world: talking to Esther Parada, Hector Méndez Caratini and Pedro Meyer," in Wombell, Paul, ed. *Photovideo.* London: Rivers Oram, 1991.

acknowledgments

If this was the very long list it should be, there would still be very special thanks to Hilda and Philip Plant, Derek Johns, Betsy Lerner, Christopher Potter, Linda Dement, Tom Epps, Nick Land, and everyone involved with Cybernetic Culture at Warwick.

All Fourth Estate books are available at your local bookshop or newsagent, or can be ordered direct from the publisher.

Indicate the number of copies required and quote the author and title.

Send cheque/eurocheque/postal order (Sterling only), made payable to Book Service by Post, to:

Fourth Estate Books
Book Service By Post
PO Box 29, Douglas
I-O-M, IM99 1BQ.

Or phone: 01624 675137

Or fax: 01624 670923

Or e-mail: bookshop@enterprise.net

Alternatively pay by Access, Visa or Mastercard

Card number: □□□□□□□□□□□□□□□□

Expiry date ....

Signature ....

Post and packing is free in the UK. Overseas customers please allow £1.00 per book for post and packing.

Name ....

Address ....

....

....

Please allow 28 days for delivery. Please tick the box if you do not wish to receive any additional information. □

Prices and availability subject to change without notice.